GEOSTATISTICAL SIMULATIONS

Quantitative Geology and Geostatistics

VOLUME 7

The titles published in this series are listed at the end of this volume.

GEOSTATISTICAL SIMULATIONS

Proceedings of
the Geostatistical Simulation Workshop,
Fontainebleau, France, 27–28 May 1993

Edited by

M. ARMSTRONG
Centre de Géostatistique,
Fontainebleau, France

and

P. A. DOWD
Department of Mining and Mineral Engineering,
University of Leeds, U.K.

Springer-Science+Business Media, B.V.

A C.I.P. Catalogue record for this book is available from the Library of Congress.

ISBN 978-90-481-4372-6 ISBN 978-94-015-8267-4 (eBook)
DOI 10.1007/978-94-015-8267-4

Printed on acid-free paper

Softcover reprint of the hardcover 1st edition 1994

Table of Contents

INTRODUCTION

When this two–day meeting was proposed, it was certainly not conceived as a celebration, much less as a party. However, on reflection, this might have been a wholly appropriate gesture because geostatistical simulation came of age this year : it is now 21 years since it was first proposed and implemented in the form of the turning bands method.

The impetus for the original development was the mining industry, principally the problems encountered in mine planning and design based on smoothed estimates which did not reflect the degree of variability and detail present in the real, mined values. The sustained period of development over recent years has been driven by hydrocarbon applications. In addition to the original turning bands method there are now at least six other established methods of geostatistical simulation.

Having reached adulthood, it is entirely appropriate that geostatistical simulation should now be subjected to an intense period of reflection and assessment. That we have now entered this period was evident in many of the papers and much of the discussion at the Fontainebleau meeting. Many questions were clearly articulated for the first time and, although many of them were not unambiguously answered, their presentation at the meeting and publication in this book will generate confirmatory studies and further research. Perhaps the most important of these questions are : How are simulation methods and simulated images to be assessed? What properties of simulated images are inherent in the methods used to obtain the images? Are some methods more suitable to specific applications than others?

The development of new methods of simulation, extensions of existing methods and new applications continues unabated as can be seen from the papers published in this volume. This continued development is driven by problems encountered in the implementation of existing methods; the search for more efficient algorithms; the desire to incorporate new variables such as (geometrical or geological) shapes; the perceived need to eliminate parametric approaches; and the new end–uses of simulated images or models.

The two–day meeting in Fontainebleau was attended by 44 persons actively engaged in research into geostatistical simulation and/or in practical applications. Whilst the attendance record was by no means an exhaustive list of those in the field, it was nevertheless a representative cross–section which included advocates of all of the established methods of, and approaches to, geostatistical simulation.

The scientific representativity of the registrants was accompanied by their geographical diversity with registratrants from Africa, Australia, North America and throughout Europe.

The reader of this volume will find a set of papers, all of which have been independently peer–reviewed and which accurately reflect the current debates, applications and research interests in geostatistical simulation. In addition, the questions and discussions which followed each paper have, as far as possible, been faithfully reproduced in the text. The organisers believe that this volume will become not just a reference and source text but will also foster and encourage further developments in what is rapidly becoming a field in its own right. As simulation begins to mature in its newly found adulthood may it lose none of the vigour, and occasional brashness, of its adolescence!

Peter Dowd
Margaret Armstrong

LIST OF PARTICIPANTS

ALLARD Denis (Mr) – Centre de Géostatistique, ENSMP
35, rue Saint–Honoré, 77305 Fontainebleau (France)
Ph: 33 1 64 69 47 87 – Fax: 33 1 64 69 47 05

ARMSTRONG Margaret (Dr) – Centre de Géostatistique, ENSMP
35, rue Saint–Honoré, 77305 Fontainebleau (France)
Ph: 33 1 64 69 47 74 – Fax: 33 1 64 69 47 05

BEUCHER Hélène (Dr) – Centre de Géostatistique, ENSMP
35, rue Saint–Honoré, 77305 Fontainebleau (France)
Ph: 33 1 64 69 47 71 – Fax: 33 1 64 69 47 05

BOURGINE Bernard (Dr) – BRGM
BP 6009, 45060 Orléans Cedex 2 (France)
Ph: 33 38 64 34 34 – Fax: 33 38 64 35 00

BREMOND Roland (Mr) – Centre de Morphologie Mathématique, ENSMP
35 rue Saint–Honoré, 77305 Fontainebleau (France)
Ph: 33 1 64 69 47 91 – Fax: 33 1 64 69 47 07

CHESSA A.G. (Dr) – Div. Mathematic Sciences – Norwegian Inst. Technology
N–7034 Trondheim (Norvège)
Ph: 47 7 ..... – Fax: 47 7 59 35 24

CHILES Jean–Paul (Dr) – BRGM
BP 6009, 45060 Orléans Cedex 2 (France)
Ph: 33 38 64 34 34 – Fax: 33 38 64 35 00

DALY Colin (Dr) – BP Research Centre, Reservoir Description
Sunbury–on–Thames TW16 7LN (GB)
Ph: 44 932 76 35 25 – Fax: 44 932 76 46 39

DERAISME Jacques (Dr) – Geovariances International
BP 91, 77212 Avon (France)
Ph: 33 1 64 22 36 15 – Fax: 33 1 64 87 28

DOWD Peter (Prof.) – Dept Mining Engineering – University of Leeds
Leeds LS2 9JT (GB)
Ph: 44 532 33 27 96 – Fax: 44 532 46 73 10

de FOUQUET Chantal (Dr) – Centre de Géostatistique, ENSMP
35, rue Saint–Honoré, 77305 Fontainebleau (France)
Ph: 33 1 64 69 47 61 – Fax: 33 1 64 69 47 05

FREULON Xavier (Dr) – 20 rue Saint–Jean
75017 Paris (France)
Ph: 33 1 42 63 75 91

FRYKMAN Peter (Dr) – Danish Geological Survey
Thoravej 8, DK–2400 Copenhagen NV (Danemark)
Ph: 45 31 10 66 00 – Fax: 45 31 19 68 68

GALLI Alain (Dr) – Centre de Géostatistique, ENSMP
35, rue Saint–Honoré, 77305 Fontainebleau (France)
Ph: 33 1 64 69 47 73 – Fax: 33 1 64 69 47 05

GEORGSEN Frode (Dr) – Norwegian Computing Centre
PO Box 117, Blindern, N–0314 Oslo 3 (Norvège)
Ph: 47 22 85 25 00 – Fax: 47 22 69 76 60

GOMEZ–HERNANDEZ José (Dr) – Depto Ingenieria Hidraulica
Universidad Politecnica de Valencia, Apartado Correos 22012,
46071 Valencia (Espagne)
Ph: 34 6 387 76 12 – Fax: 34 6 387 76 18

GOTWAY Carol (Dr) – Dept of Biometry – University of Nebraska
PO Box 830712, Lincoln, Ne 68583–0712 (USA)
Ph: 1 402 472 29 03 – Fax: 1 402 472 56 39

HAAS André (Dr) – Elf Aquitaine Production, PRO/RDP
BP 65, 64018 Pau Cedex (France)
Ph: 33 59 83 40 00 – Fax: 33 59 83 49 80

HEGSTAD Bjorn (Dr) – Div. Mathematic Sciences – Norwegian Inst. Technology
N–7034 Trondheim (Norvège)
Ph: ..... – Fax: 47 7 59 35 24

HEKTOEN Anne–Lise (Dr) – Norwegian Computing Centre
PO Box 117, Blindern, N–0314 Oslo 3 (Norvège)
Ph: 47 22 ..... – Fax: 47 22 69 76 60

HU Linying (Dr) – Centre de Géostatistique, ENSMP
35, rue Saint–Honoré, 77305 Fontainebleau (France)
Ph: 33 1 64 69 47 65 – Fax: 33 1 64 69 47 05

HUDSON Gordon (Mr) – Macaulay Land Use Research Inst.
Craigiebuckler, Aberdeen AB9 2QJ (GB)
Ph: 44 224 31 86 11 – Fax: 44 224 31 15 56

JAQUET Olivier (Mr) – Colenco
Mellingerstrasse 207, 5405 Baden (Suisse)
Ph: 41 56 77 15 76 – Fax: 41 56 83 73 57

JESEK Josef (Dr) – Faculty of Science
Albertov 6, 12843 Praha 2 (Tchécoslovaquie)

JEULIN Dominique (Dr) – Centre de Géostatistique, ENSMP
35, rue Saint–Honoré, 77305 Fontainebleau (France)
Ph: 33 1 64 69 47 95 – Fax: 33 1 64 69 47 05

KLEINGELD Wynand (Dr) – Ore Evaluation Dpt – AAC/De Beers Ltd
PO Box 41848, Craighall 2024 (Afrique du Sud)
Ph: 27 11 880 68 80 – Fax: 27 11 880 62 98

LAJAUNIE Christian (Dr) – Centre de Géostatistique, ENSMP
35, rue Saint–Honoré, 77305 Fontainebleau (France)
Ph: 33 1 64 69 47 63 – Fax: 33 1 64 69 47 05

LANGLAIS Valérie (Dr) – Dept Reservoir Technology – Statoil
PO Box 300, N–4001 Stavanger (Norvège)
Ph: 47 4 80 78 96 – Fax: 47 4 80 96 72

LANTUEJOUL Christian (Dr) – Centre de Géostatistique, ENSMP
35, rue Saint–Honoré, 77305 Fontainebleau (France)
Ph: 33 1 64 69 47 66 – Fax: 33 1 64 69 47 05

LE LOC'H Gaëlle (Dr) – Centre de Géostatistique, ENSMP
35, rue Saint–Honoré, 77305 Fontainebleau (France)
Ph: 33 1 64 69 47 72 – Fax: 33 1 64 69 47 05

MASTERS Stuart (Mr) Minenco Pty Ltd, GPO Box 5202AA, Melbourne 3001, Australia
Ph: 61 3695 4409 – Fax: 61 3695 4362

MOSTAD Petter (Dr) – Norwegian Computing Centre
PO Box 117, Blindern, N–0314 Oslo 3 (Norvège)
Ph: 47 22 ..... – Fax: 47 22 69 76 60

OMRE Henning (Dr) – Div. Mathematic Sciences – Norwegian Inst. Technology
N–7034 Trondheim (Norvège)
Ph: 47 7 ..... – Fax: 47 7 59 35 24

PANDALAI Hari (Dr) – Indian Institute of Technology – Dpt of Earth Sciences
Powai, Bombay 400 076 (Inde)
Ph: 91 22 578 25 45 – Fax: 91 22 578 34 80

RAVENSCROFT Peter (Mr) – RTZ Technical Services Ltd
PO Box 50, Castlemead, Bristol BS99 7YR (GB)
Ph: 44 272 27 ..... – Fax: 44 272 27 33 17

RENARD Didier (Mr) – Centre de Géostatistique, ENSMP
35, rue Saint–Honoré, 77305 Fontainebleau (France)
Ph: 33 1 64 69 47 80 – Fax: 33 1 64 69 47 05

RIVOIRARD Jacques (Dr) – Centre de Géostatistique, ENSMP
35, rue Saint–Honoré, 77305 Fontainebleau (France)
Ph: 33 1 64 69 47 64 – Fax: 33 1 64 69 47 05

RUFFO Paolo (Dr) – AGIP – Elab. Dept
PO Box 12069 – 20120 Milano (Italy)
Ph: 39 2 520 43 65 – Fax: 39 2 520 22 127

RUTHERFORD Brian (Dr) – Sandia National Labs nº 323
Albuquerque, New Mexico 87185 (USA)
Ph: 1 505 844 31 20 – Fax: 1 505 844 90 37

SAMPSON Paul (Dr) – University of Washington – Dept of Statistics GN–22
Seattle, Wa 98195 (USA)

SEGURET Serge (Dr) – Centre de Géostatistique, ENSMP
35, rue Saint–Honoré, 77305 Fontainebleau (France)
Ph: 33 1 64 69 47 85 – Fax: 33 1 64 69 47 05

SMITHIES Steve (Mr) – Genmin Geostats Dept
PO Box 61820, Marshalltown 2107 (Afrique du Sud)
Ph: 27 11 376 32 46 – Fax: 27 11 376 23 92

TJELMLAND Hakom (Dr) – Div. Mathematic Sciences – Norwegian Inst. Technology
N–7034 Trondheim (Norvège)
Ph: 47 7 ..... – Fax: 47 7 59 35 24

WACKERNAGEL Hans (Dr) – Centre de Géostatistique, ENSMP
35, rue Saint–Honoré, 77305 Fontainebleau (France)
Ph: 33 1 64 69 47 60 – Fax: 33 1 64 69 47 05

STOCHASTIC SIMULTATION FOR IMAGING SPATIAL UNCERTAINTY: COMPARISON AND EVALUATION OF AVAILABLE ALGORITHMS

C. A. GOTWAY
Department of Biometry
University of Nebraska
Lincoln, NE 68583-0712

B.M. RUTHERFORD
Sandia National Laboratories
P.O Box 5800
Albuquerque, NM 87185

Abstract. Stochastic simulation has been suggested as a viable method for characterizing the uncertainty associated with the prediction of a nonlinear function of a spatially-varying parameter. Geostatistical simulation algorithms generate realizations of a random field with specified statistical and geostatistical properties. A nonlinear function (called a transfer function) is evaluated over each realization to obtain an uncertainty distribution of a system response that reflects the spatial variability and uncertainty in the parameter. Crucial management decisions, such as potential regulatory compliance of proposed nuclear waste facilities and optimal allocation of resources in environmental remediation, are based on the resulting system response uncertainty distribution.

Many geostatistical simulation algorithms have been developed to generate the random fields, and each algorithm will produce fields with different statistical properties. These different properties will result in different distributions for system response, and potentially, different managerial decisions. The statistical properties of the resulting system response distributions are not completely understood, nor is the ability of the various algorithms to generate response distributions that adequately reflect the associated uncertainty.

This paper reviews several of the algorithms available for generating random fields. Algorithms are compared in a designed experiment using seven exhaustive data sets with different statistical and geostatistical properties. For each exhaustive data set, a number of realizations (both unconditional and data-conditioned) are generated using each simulation algorithm. The realizations are used with each of several deterministic transfer functions to produce a cumulative uncertainty distribution function of a system response. The uncertainty distributions are then compared to the single value obtained from the corresponding exhaustive data set. The results of the study facilitate comparisons between the individual methods, allow an assessment of the consistency of the simulation algorithms, and indicate potential for bias or imprecision.

INTRODUCTION

Stochastic simulation provides a way to incorporate various types of uncertainty into prediction of a complex system response. Usually, some information is available on a parameter of interest (for example, the permeability of a sandstone formation), but the transfer function (a groundwater flow model, for example) may require a detailed spatial map of this parameter. The exhaustive sampling necessary to obtain such a map is usually not feasible. One alternative is to generate realizations of a

M. Armstrong and P. A. Dowd (eds.), Geostatistical Simulations, 1–21.

random field that share the available information on the parameter of interest. These realizations serve as input to the transfer function that computes a system response for each. If the realizations characterize the spatial uncertainty of the parameter of interest, the resulting distribution of predicted system response values will reflect the uncertainty (see Figure 1). This approach, proposed in Journel (1989), is widely used in hydrology, petroleum engineering, and the environmental sciences. Crucial management decisions, such as potential regulatory compliance of proposed nuclear waste sites and optimal allocation of resources in environmental remediation, are based on the resulting system response uncertainty distribution.

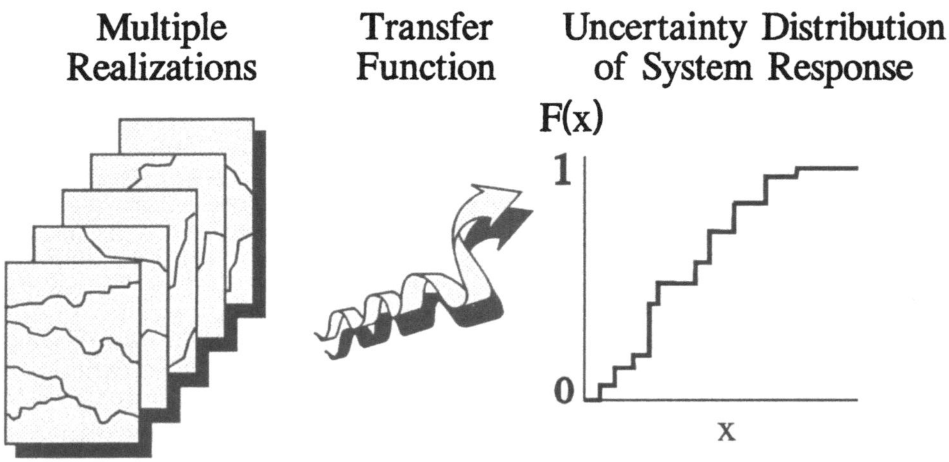

Figure 1. Schematic Illustration Of Stochastic Simulation. Modified from Journel (1989)

Many different stochastic simulation algorithms could be used to construct the realizations and each may assign different statistical and spatial features to the generated fields. These differences may be due to the order in which the simulated values are obtained (the method of generation), the statistical and geostatistical inputs to the algorithms (i.e., the summary functions such as the variogram), and the degree to which the summary functions of the realizations match the specified input summary functions. Consequently, the distribution of predicted response values, and the resulting inferences drawn from this distribution, depend on the particular simulation algorithm implemented. For the scientist using stochastic simulation, the differences among the algorithms and the resulting effects on the system response uncertainty distributions need to be understood so that appropriate algorithms are used in each application.

Very little has been done to evaluate and compare these algorithms as they are currently being used. Journel and Alabert (1989) use the Berea Sandstone exhaustive data set to compare sequential indicator simulation to Gaussian simulation. They recommend the use of indicator methods when it is necessary to characterize strong connectivity of extreme (high or low) values. Deutsch and Journel (1992b) use the same data set to compare sequential indicator simulation, sequential Gaussian simulation, and simulated annealing using petroleum industry-based transfer functions. Based on results from one set of simulated system response distributions, they find all methods to be feasible, accurate, and precise. Hansen (1992) uses several indicator-based synthetic exhaustive data sets to evaluate the sequential indicator simulation algorithm. The results of this study indicate that the sequential indicator simulation algorithm, when applied to problems where conditioning data are available, may "over-condition" the data, causing the response uncertainty distribution to be very precise but biased for the true response as obtained from the exhaustive data sets. Clearly, much more work is needed to evaluate the many geostatistical simulation algorithms over the range of applications considered in practice.

The purpose of this paper is to present the results of a comprehensive study designed to evaluate and compare geostatistical simulation algorithms using a number of different exhaustive data sets that represent a variety of spatial phenomena. The objective of the study, discussed in subsequent sections, is to obtain more information on how the different simulation algorithms work in basic simulation applications. Using them in a designed experiment with many exhaustive data sets will facilitate comparison of realizations and response uncertainty distributions produced by the various simulation algorithms, allow an assessment of the robustness of the methods to differences in underlying data distributions, and indicate potential biases or imprecision due to a particular algorithm or method. Section II gives the specifics of the design of this experiment, and Section III summarizes the results. A discussion of the findings and recommendations for implementation and future research are then given in Section IV.

DESIGN OF THE COMPARATIVE EXPERIMENT

The basic experiment is as follows. For each of seven exhaustive data sets, 200 realizations of a random field model are generated using the simulation algorithms appropriate for that data set. Input to each simulation method is based on "truth" as derived from the exhaustive data set. Truth will be quantified using appropriate "summary functions," i.e., functions that incorporate the statistical and geostatistical features of the field, such as the semivariogram or indicator semivariogram. One-hundred of the fields are conditioned on data randomly selected from the original exhaustive data set. The remaining 100 fields are left unchanged as unconditional realizations. For each realization, several transfer functions are computed, each giving one value of the system response. The results of the experiment provide several uncertainty distributions for every simulation algorithm, each of which corresponds to one of the exhaustive-data-set/transfer-function scenarios. Half of these distributions will be based on unconditional realizations, and the other half will be data-conditioned.

A. *Simulation Algorithms and Software*

A brief description of all algorithms and software used in this study is provided below. Because of time and space constraints, we have omitted several promising simulation algorithms such as fractal simulation algorithms, spectral algorithms, Boolean and random set algorithms, and nearest-neighbor approaches. These methods will be explored in subsequent studies.

1. LU Decomposition. The LU decomposition method is based on an LU or Cholesky-type of decomposition of the covariance matrix between data locations

and grid locations. Specifically, this covariance matrix can be decomposed as

$$C = \begin{pmatrix} C_{11} & C_{12} \\ C_{21} & C_{22} \end{pmatrix} = LU = \begin{pmatrix} L_{11} & 0 \\ L_{21} & L_{22} \end{pmatrix} \begin{pmatrix} U_{11} & U_{12} \\ 0 & U_{22} \end{pmatrix},$$

where C_{11} is the covariance between data at data locations, C_{22} is the covariance between data at grid locations, and C_{12} is the covariance between data at data locations and those at grid locations. A conditional Gaussian simulation is obtained by simulating a vector ϵ of independent normal random variables with mean zero and unit variance, and using the data vector z in the transformation

$$\begin{pmatrix} L_{11} & 0 \\ L_{21} & L_{22} \end{pmatrix} \begin{pmatrix} L_{11}^{-1} z \\ \epsilon \end{pmatrix} = \begin{pmatrix} z \\ L_{21} L_{11}^{-1} z + L_{22}\epsilon \end{pmatrix}.$$

Further details of this algorithm can be found in Davis (1987a), Cressie (1991), and Dowd (1992). LU decomposition is relatively easy to implement, can handle any type of covariance function and anisotropy, and can incorporate data conditioning efficiently. However, the amount of storage required can limit the size of the simulation grid that can be efficiently considered. Moreover, when the simulation grid size is large and the covariance matrix is sparse, numerical inaccuracies may result. To circumvent these difficulties, approximations that provide more efficient and stable calculations have been introduced by Quimby (1986) and Davis (1987b).

The basic LU decomposition algorithm in Deutsch and Journel (1992a) was used in generating the continuous-variable simulations. The random number generator provided in the software was replaced with one found in Press, et al. (1986). In implementing the simulations required for this study, we found storage space and run times to be more of a nuisance than a limitation. Each conditional simulation took about 16 minutes on an IBM RS6000 workstation, which given the computational nature of many environmental applications, is hardly a limitation.

2. Turning Bands. The turning bands method was developed to ease the computational burden in generating three dimensional fields. The method works by simulating one-dimensional processes on lines regularly spaced in two- or three-dimensions. The one-dimensional simulations are then projected onto the spatial coordinates and averaged to give the required two- or three-dimensional simulated value. The turning bands algorithm is a fast and efficient method of random field generation, but the use of a separate data-conditioning step based on kriging can reduce its efficiency for generating conditional random fields. Perhaps the biggest drawback of the method is the limitation on the choice of covariance function that can be specified. One list of possible choices is provided in Zimmerman and Wilson (1990), and additional descriptions and properties of this algorithm can be found in Journel (1974), and Mantoglou and Wilson (1982).

The turning bands computer code TUBA (Zimmerman and Wilson, 1990) was used to generate the continuous-variable simulations since this code provides a high degree of flexibility in the choice of turning bands parameters. To reduce banding artifacts due to the one-dimensional line processes, 64 turning bands were used in every simulation at the recommendation of the author of the TUBA code. The software allows the user several choices for the covariance function, but it does not generate fields with a specified nugget effect. It does not (nor does any other turning bands code known to the authors) easily incorporate zonal anisotropy or anisotropy that departs from the coordinate directions. To use the turning bands code in such situations requires the addition of two or more realizations with simple covariance structures. These additional computations were done for simulations in which a nugget effect was required, but, in the cases where complex semivariogram models were necessary, approximate models, compatible with the turning bands software, were used.

3. Sequential Gaussian and 4. Sequential Indicator Simulation Algorithms. Both of these methods are based on a sequential approach to simulation described in Journel and Alabert (1989), Gomez-Hernandez and Srivastava (1990), and Deutsch and Journel (1992a). The basic conditional sequential simulation algorithm is as follows: 1) Define a random path through all grid nodes; 2) Draw a value from the conditional distribution of the random variable at the first grid node given the (n) conditioning data; 3) Add this new value to the conditioning data set; 4) Draw a value from the conditional distribution of the variable at node two given the (n+1) conditioning data, and 5) Repeat until all nodes are simulated. In sequential Gaussian simulation, the conditioning data are first transformed to standard Gaussian values and the semivariogram of the transformed data is specified. Simple kriging is used to obtain estimates of the necessary conditional distributions. At each node, the kriged value obtained from simulated and conditioning data, and the associated kriging variance are used to specify the conditional Gaussian distribution. Realizations are then drawn randomly from this distribution. Finally, the results of the Gaussian simulation are transformed back to the original data space. In the sequential indicator simulation approach, no assumptions are made about the parametric form of the conditional distributions. The conditioning data are transformed to indicators defined by threshold values based on available data and other relevant information. Estimates of the conditional distributions at each grid node are given by simple indicator kriging using corresponding indicator semivariograms.

As discussed in Dowd (1992) these methods have several advantages including automatic handling of anisotropies and data conditioning, and fast computer implementation since an efficient kriging algorithm with a moving neighborhood search capability is all that is required. However, since sequential methods are relatively new, their properties and limitations, if any, are unknown. Artifacts of these algorithms could be present in the generated fields or in the system response uncertainty distribution. In particular, the conditional distributions obtained from the sequential indicator simulation algorithm do not respect the properties of cumulative distribution functions. It is possible to obtain probability estimates larger than 1, less than zero, and often the resulting conditional distributions are not monotonic. Although an artificial correction is used to force the desired properties, it is not clear what effect this correction may have on the realizations and the system response distribution.

Both sequential algorithms were used for all continuous-variable simulations using the software given in Deutsch and Journel (1992a). As with LU decomposition, the random number generator was replaced by that given in Press, et al., (1986). Both algorithms were very flexible, efficient, and easy to use. However, several user-specified parameters– such as the use of simple kriging versus ordinary kriging, the maximum number of simulated nodes retained for kriging, octant-search parameters, and, in particular, upper and lower tail extrapolation choices– can affect the efficiency of the algorithms, the nature of the realizations, and the resulting uncertainty distributions.

Sequential indicator simulation is especially straightforward for generating realizations of a categorical variable. Thus, this algorithm was also used to produce realizations of the GCD and Boolean exhaustive data sets (discussed below).

5. Truncated Gaussian Random Function Approach This method, discussed in Matheron et al., (1987), Galli et al., (1990), and Dowd (1992) describes K lithofacies or lithologic units by using one indicator function per facies. Indicator simulation is accomplished by generating a Gaussian random field and then truncating the Gaussian values to achieve the categorical simulation. Specifically, suppose $Y(s)$ is a standard Gaussian random variable. The associated indicator transform is

$$I(s;y_i) = \begin{cases} 1 & \text{if } Y(s) \in (y_{i-1}, y_i] \\ 0 & \text{otherwise} \end{cases}$$

so that a point s belongs to category i if $Y(s) \in (y_{i-1}, y_i]$. The thresholds, y_i, are determined according to the proportion of values that fall into each category. Let Φ be the standard Gaussian distribution function, and let p_i be the proportion of values in category i. Then

$$p_1 \equiv P[-\infty < Y(s) \leq y_1] \quad \rightarrow y_1 = \Phi^{-1}(p_1).$$

In general, $y_{k-1} = \Phi^{-1}(p_1 + p_2 + \ldots + p_{k-1})$. After the Gaussian thresholds have been determined, it is then necessary to determine the covariance structure of the Gaussian random variables. This structure depends on the structure of the indicator covariances which may be computed and modeled from the data. The relationship between the two covariance functions can be specified using Hermite polynomial expansions, in which case (Dowd, 1992)

$$C_I(h) = \sum_{i=1}^{K}\sum_{j=1}^{K} c_i c_j g(y_i)g(y_j) + \sum_{n=1}^{\infty} \frac{H_{n-1}(y_i)H_{n-1}(y_j)}{n!}(C_Y(h))^n,$$

where $H_{n-1}(x)$ are Hermite polynomials, g(y) is the standard Gaussian density, c_i is a unique integer value assigned to each category, and $C_I(h)$, and $C_Y(h)$ are the covariance functions of the indicator variables and the Gaussian variables, respectively. Conditioning data at locations $\{s_\alpha\}$, are replaced by standard Gaussian random variables with covariance function $C_Y(h)$ such that $y_{j-1} \leq y(s_\alpha) < y_j$, where j indexes the category to which s_α belongs. Finally, a usual conditional Gaussian simulation is performed and then back-transformed to obtain the associated indicator values.

There are many ways to determine $C_Y(h)$ from the above equation. To the authors' knowledge, there is no published theory on an optimal solution for $C_Y(h)$ or even an accepted methodolgy for obtaining any satisfactory solution. The approach used in this study was to obtain values of $C_Y(h)$ that satisfy the above equation using a Golden Section Search algorithm (Press, et al., 1986) for all relevant lags h, rather than to use a specific parametric model. Once $C_Y(h)$ has been established, Gaussian random fields were generated via simulated annealing (Kirkpatrick, et al., 1983; Deutsch and Journel 1992a, 1992b) to force each realization to match $C_Y(h)$. The determination of $C_Y(h)$ is the most computationally intensive aspect of this algorithm, and the lack of a straightforward method for conditioning to indicator data is also a limitation. For problems with more than two categories, consideration of indicator cross covariances is required, and additional computations are necessary.

Clearly, in this type of study, it is impossible to divorce the simulation method or algorithm from the computer software used to implement the algorithm. In constructing the software, there are often various choices for the computational details of algorithm implementation (e.g., the number of turning bands lines, cdf interpolation and extrapolation methods, search strategies) that are specified by the user. Every attempt was made to provide the most complete input to each simulation code. The goal of this study is not to discredit any of the approaches or to determine the best algorithm, since no algorithm can be best for all applications. Instead, we hope to obtain more information on how the algorithms work in simulation applications like those illustrated in Figure 1 and to indicate new research directions for investigating the properties of the algorithms and developing new algorithms.

B. *Data conditioning*

For the conditional realizations, N=100 sets of r=100 data values were randomly selected from each exhaustive data set. Each realization was then conditioned on one of the N sets of 100 values. Although in practice all realizations are conditioned to the same set of data values, in a study such as this, care must be taken not to introduce bias into the experiment due to the particular data set chosen. Since it is impossible to determine a "good" data set *apriori,* multiple sets of conditioning data were used.

C. *Exhaustive Data Sets*

Seven exhaustive data sets were used in this comparative study. Three are real exhaustive data sets taken from the authors' project work and from the literature. The advantage of these data sets is that they reflect real spatial phenomema and the statistical and geostatistical properties of these data sets reflect what is actually encountered in practice. However, results based on real exhaustive data sets are limited to the particular properties exhibited by the data. In order to broaden the scope of inference of a study such as the one described here, it is necessary to consider exhaustive data sets with a variety of statistical and geostatistical properties. Thus, the remaining four exhaustive data sets were generated synthetically so that their statistical and geostatistical features could be controlled and varied. A brief description of each exhaustive data set is provided below and corresponding gray-scale maps are shown in Figure 2.

1. Multivariate Gaussian Exhaustive Data Set. This data set was obtained by generating 1600 independent standard Gaussian random variables on a 40 x 40 grid of unit spacing. The covariance matrix was derived from a prespecified exponential semivariogram model with a zero nugget, unit sill, and range 7. A Cholesky decomposition algorithm from the IML procedure in SAS (Statistical Analysis System Version 6.01, 1992) was used to induce the desired covariance structure. The ensemble average was then standardized to mean zero and unit variance to produce the exhaustive data set.

2. Uniform Exhaustive. The uniform exhaustive data set consists of 1600 values on a 40 x 40 regular grid with unit spacing and was obtained by generating a set of uniform random variables over the grid and then applying simulated annealing to more closely match the desired spatial covariance. The simulated annealing algorithm given in GSLIB (Deutsch and Journel, 1992a) was used to force the image to match (to the extent permitted by this algorithm) two summary functions: 1) the histogram of a uniform probability distribution on $[-\sqrt{3},\sqrt{3}]$, and 2) an isotropic exponential semivariogram with zero nugget, unit sill, and range 10.

3. Indicator-Covariance Specified Exhaustive. This exhaustive data set was obtained by modifying the simulated annealing algorithm given in Deutsch and Journel (1992a) to anneal an initial image to specified indicator semivariograms. The resulting exhaustive data set was forced to have mean zero, unit variance, and indicator semivariograms specified as
Cut-off -2: isotropic sph ($c_o = 0.00, c_s = 0.16, a = 7.00$),
Cut-off -1: isotropic sph ($c_o = 0.00, c_s = 0.25, a = 5.00$),
Cut-off 0: anisotropic sph ($c_o = 0.00, c_s = 1.00, ax = 7.00, ay = 3.50$),
Cut-off 1: isotropic sph ($c_o = 0.00, c_s = 1.25, a = 10.0$),
Cut-off 2: anisotropic sph ($c_o = 0.00, c_s = 1.50, ax = 30.00, ay = 15.00$).
The notation "sph" refers to a spherical semivariogram model with nugget c_o, sill c_s, and range a. The notation "anisotropic sph" specifies a model of geometric anisotropy, where the parametric form is that of a spherical semivariogram model, but with a range of ax in the N 90° E direction and a range of ay in the N 0° E direction.

4. Reduced Walker Lake Exhaustive. To construct the exhaustive data set called the "Reduced Walker Lake" exhaustive data set, an initial exhaustive variable, U', was selected from a portion of the U variable of the original Walker Lake Data Set (Isaaks and Srivastava, 1989). The reduction was done in order to facilitate comparisions with results obtained for other continuous-variable exhaustive data sets that consist of 1600 data points on a regular 40 x 40 grid with unit spacing. Following the approach of Desbarats and Srivastava (1991), the variable U' was then transformed to create a variable T which is similar to the log-normally distributed transmissivity parameter common in hydrogeologic applications.

5. Berea Sandstone Exhaustive. The Berea sandstone data set consists of 1600 air permeameter measurements (in millidarcies) taken from a 2 x 2 foot vertical slab of Berea sandstone (Giordano, et al., 1985). The locations of these measurements are equally spaced over a 40 x 40 grid. The Berea sandstone has been used in many comparative geostatistical studies such as Journel and Alabert (1989), Deutsch and Journel (1992b), and Rossi and Posa (1992). Figure 2e shows the the banding in the N 123°E direction that is characteristic of this exhaustive data set.

6. GCD Exhaustive. The GCD exhaustive data set presented in this manuscript is a coarser version of the one discussed in Gotway, et al. (1993) obtained by digitizing a photo-mosaic map of a trench wall at the Greater Confinement Disposal Facility in Southern Nevada. Each node on the resulting grid is associated with one of two hydrogeologic units and is assigned a value of 1 (for the black unit) or 0 (for the white unit). For the purposes of this paper, the exact nature of each unit is not important.

7. Boolean Exhaustive. This exhaustive data set was obtained by Boolean simulation of half-ellipses. Ellipses were horizontally oriented with anisotropy ratios of 2 or 3 and minor axes lengths varying from 1-5 units. The target coverage percentage was specified at 20 percent. Each node on the 80 x 23 grid was given a value of 1 if it was contained in at least one of the ellipses and the value 0 if it was not contained in any ellipse.

Summary functions necessary for input into the various simulation algorithms were obtained directly from the exhaustive data sets. For the Gaussian-based simulation algorithms, the continuous, non-Gaussian, exhaustive data sets were transformed to normality prior to simulation. Semivariograms (N 0°E, N 45°E, N 90°E, N 135°E, and omnidirectional), were then determined from the transformed

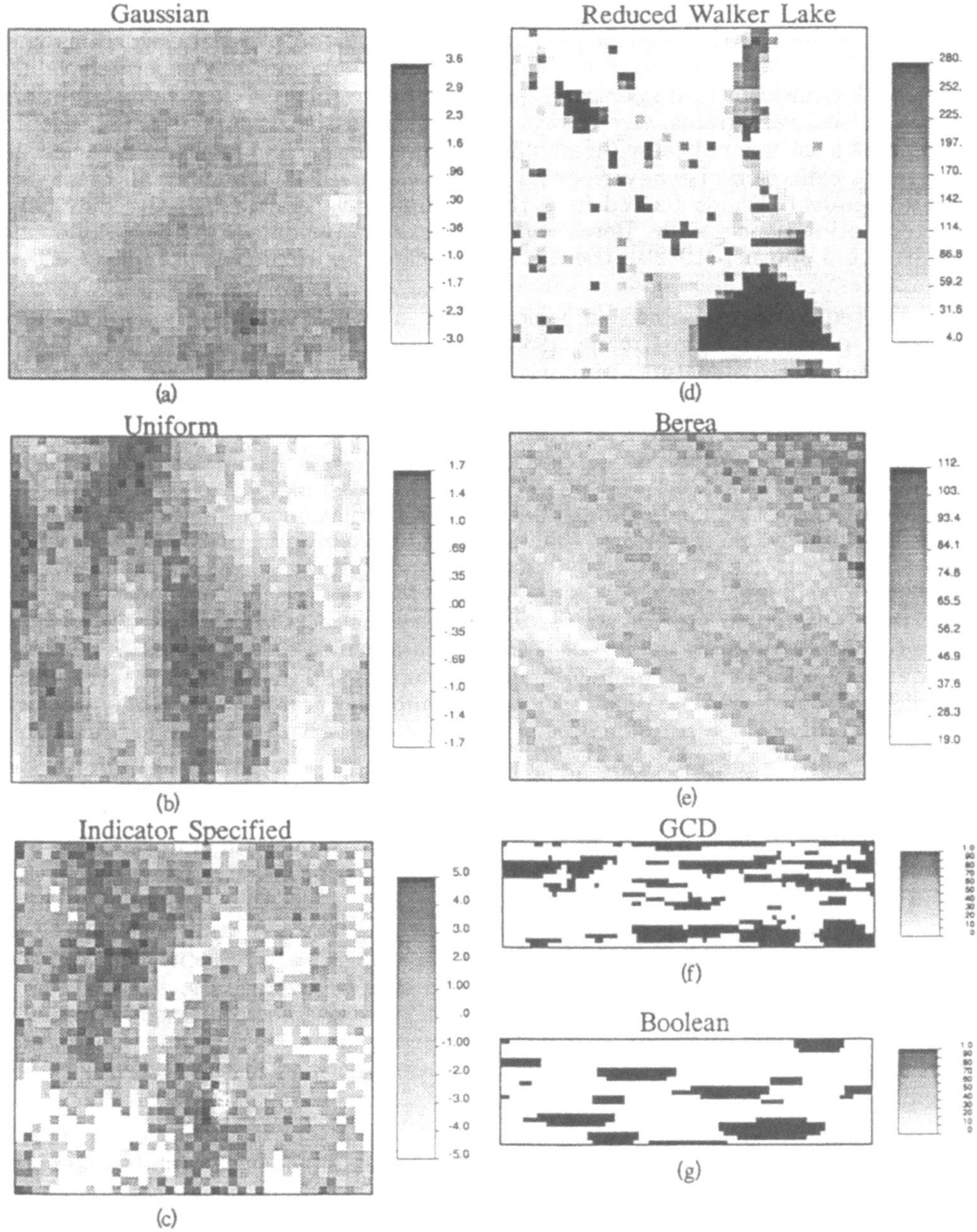

Figure 2. Exhaustive Data Sets.

data and used to generate the realizations. For the real data sets where additional information indicated the principal directions of anisotropy, semivariograms corresponding to these directions were used. After simulation, the generated values were back-transformed, if necessary, based on the original exhaustive distribution. Indicator semivariograms were always obtained from the original exhaustive data set and at least five indicator thresholds were used for each exhaustive data set. In general, a collection of semivariogram models was used in modeling all of the indicator semivariograms derived from the continuous-variable exhaustive data sets. The exception concerns the Berea sandstone exhaustive data set where, following Deutsch and Journel (1992b), the median-indicator approximation (Journel, 1983) was used.

All semivariograms were modeled using a combination of weighted-least-squares-regression fitting (Cressie, 1985) and visual fitting techniques. Anisotropic models and nested structures were used where appropriate. Additional constraints were placed on the choice of indicator semivariogram models in order to respect their theoretical properties.

D. *Transfer functions*

Many important applications of stochastic simulation are flow related. Measures of groundwater travel time, contaminant breakthrough time, and bed thickness all rely on some quantitative notion of connectivity that reflects the degree to which regions of high or low values are related. Many of the transfer functions used here were selected to provide simple indicators of connectivity. They are used in place of actual groundwater flow or transport codes since the complexities involved in boundary condition determination and assumptions, potential errors introduced by necessary grid-discretization, and model calibration issues would only detract from the ability to detect differences in statistical properties among the simulation algorithms.

There are other important transfer functions that are not necessarily flow related. For example, the ability to accurately predict the proportion of values above a specified threshold is important in mining and environmental restoration applications. In these and many other applications, incorporation of major geologic features may also be a primary concern.

The transfer functions described below were selected to encompass a variety of applications like those described above. For the categorical problems, the transfer functions are called discrete path, cluster, and discrete average cost. These functions operate on binary data sets. For the continuous-variable problems, the transfer functions are called minimum cost path, geometric mean average and range, and threshold proportion. The transfer functions are described in the following paragraphs.

<u>Discrete Path.</u> At each x-node along the upper boundary of the data set, a particle is released. The particle can move downward or diagonally downward, but if no path is available it can move only horizontally. Barriers to movement are considered to be the higher coded (black) materials in both problems considered here. The initial direction of horizontal movement is left to right. The direction is changed whenever a barrier is encountered during an attempted horizontal move. The output from the transfer function is the number of particles reaching the lower boundary.

<u>Cluster.</u> This algorithm counts the number of clusters of barrier material. The barrier materials are considered to be part of the same cluster if they are connected diagonally, horizontally, or vertically.

Discrete Average Cost. Particles are released from each node on the left boundary and travel horizontally through the region with unit penalty for each move within a white region. Particles can move right or diagonally right to avoid the black regions where movement costs are five times as great. Output is the average cost.

Minimum-Cost Path. In this algorithm, movement costs are based on the reciprocals of the data values. The minimum cost path from the upper boundary to the lower boundary is computed using a dynamic programming algorithm allowing downward and diagonally-downward movement. The output is the minimum cost.

Geometric Mean. The geometric mean is computed for each interior grid node as the product of the 25 closest nodes. For data sets with negative measurements, a constant was added to assure all values were positive. A second fixed constant was used to control the magnitude of the results. Two transfer functions were considered: the difference between the maximum and minimum of all the geometric means; and the average of all geometric means.

Threshold Proportion. This transfer function is the proportion of values that are greater than the 90th percentile of the exhaustive data set.

The transfer functions are applied first to the exhaustive data sets to obtain the "true" value to be used as a basis for comparison, and then to each realization. Comparisons between these values are then used in Section III to draw inferences concerning the simulation techniqies.

RESULTS

The results of this study reveal a number of differences between the simulation algorithms. The differences can be seen in 1) the accuracy of the uncertainty distributions, 2) the characteristics of the realizations input to the transfer functions, and 3) the shapes of the system response distributions. Results in each of these areas are summarized and discussed below.

To help assess the methods, each system response uncertainty distribution was compared to the true value computed from the exhaustive data set. Table 1 gives a summary of the overall results. Columns one and two provide the proportion of system response distributions that 1) contained the true value within the range of response values, and 2) contained the true value within the 90th and 10th percentiles. The last three columns of Table 1 provide additional measures pertaining to the bias, precision, and accuracy of the response uncertainty distributions produced by each algorithm. The bias measure (column 3) is the absolute difference between the median of the uncertainty distribution and the true value, divided by the true value, then averaged over all exhaustive-data set/transfer-function combinations. Precision (column 4) is measured as the difference between the 90th and 10th percentiles of each uncertainty distribution divided by the corresponding percentile difference for the uncertainty distributions obtained using the unconditional LU decomposition simulation algorithm or the unconditional categorical sequential indicator algorithm (as arbitrary references), averaged over all exhaustive-data-set/transfer-function combinations. Simulation algorithms that consistently produce uncertainty distributions that are more precise than those obtained using unconditional LU decomposition (or categorical sequential indicator) simulation will have precision values less than 1.00. Column 5 combines the measures of bias and precision into a measure of accuracy. Accuracy is measured as the weighted average of the absolute differences between system response value and true value, where the weights are the reciprocals of the across-simulation-method average standard deviation of the uncertainty distributions for each exhaustive-data-set/transfer-function scenario. The weights are used to account for differences in the magnitude of the values in the different scenarios. Lower numbers for the accuracy measure are indicative of uncertainty distributions whose values are consistently close to those computed from the exhaustive data sets.

SIMULATION ALGORITHM	PROPORTION OF DISTRIBUTIONS CONTAINING THE TRUE VALUE OF $(X_{(min)}, X_{(max)})$		PROPORTION OF DISTRIBUTIONS CONTAINING THE TRUE VALUE OF $(X_{.10}, X_{.90})$		BIAS		PRECISION		ACCURACY	
	Unc.	Cond.	Unc.	Cond.	Unc.	Cond.	Unc.	Cond.	Unc.	Cond.
TURNING BANDS	18/20	19/20	17/20	18/20	0.35	0.18	0.71	.45	1.14	1.52
LU DECOMPOSITION	20/20	19/20	19/20	17/20	0.30	0.17	1.00	0.45	1.25	1.47
SEQUENTIAL GAUSSIAN	20/20	18/20	18/20	17/20	0.31	0.17	0.71	0.37	1.04	1.52
SEQUENTIAL INDICATOR (continuous)	17/20	17/20	13/20	14/20	0.51	0.32	0.88	0.37	1.17	1.75
SEQUENTIAL INDICATOR (categorical)	5/6	5/6	5/6	5/6	0.35	0.28	1.00	.91	1.33	1.32
TRUNCATED GAUSSIAN	4/6	4/6	3/6	3/6	0.43	0.42	.90	.93	1.64	1.72

Table 1. Summary of Uncertainty Distributions Combined Across All Exhaustive-Data Set/Transfer-Function Scenarios. The bias, precision, and accuracy measures are those discussed in the text.

The first column of Table 1 shows that most of the uncertainty distributions produced by the continuous-variable simulation algorithms did contain the true value. The exceptions are the uncertainty distributions based on transfer functions used with the Reduced Walker Lake exhaustive data set. The second column of Table 1 gives the proportion of uncertainty distributions containing the true value within the 10th and 90th percentiles. On average, this number should be around 80%, or 16/20, for the continuous-variable simulation algorithms. Thus, from the table we can see that the probability content assigned to specified intervals appears to be accurate for uncertainty distributions based on conditional Gaussian algorithms, but may be too low for uncertainty distributions produced using the sequential indicator algorithm, or too high if unconditional LU decomposition or sequential Gaussian methods are used for problems of the type presented here.

The last three columns of Table 1 show that the uncertainty distributions obtained using the unconditional Gaussian-based simulation algorithms have similar values for the bias measure, but that those obtained using the unconditional turning bands and sequential Gaussian algorithms are more precise than those based on the unconditional LU decomposition algorithm. Conditioning the simulation reduces the differences in the uncertainty distributions produced by the various methods, but, on average, uncertainty distributions produced by the Gaussian-based simulation algorithms are more accurate than those produced using sequential indicator simulation. These results differ from those presented by Journel and Deutsch (1993) in which response uncertainty distributions based on unconditional realizations generated by the sequential Gaussian simulation algorithm were compared to those obtained using sequential indicator simulation. Based on one uncertainty distribution that did not contain the true value, Journel and Deutsch (1993) conclude that Gaussian random field models may produce uncertainty distributions and probability intervals that are too narrow. Clearly, the results of our study do not support their contention. Although the Gaussian realizations produced an inaccurate response uncertainty distribution for the exhaustive-data-set/transfer function scenario considered in Journel and Deutsch (1993), this may be due to the use of the sequential Gaussian simulation algorithm, and conclusions based on this algorithm cannot be extended to Gaussian-based methods in general.

Most of the response uncertainty distributions obtained using the categorical simulation algorithms were also accurate. The primary exceptions are the results for the Boolean exhaustive data set and the discrete path transfer function, where none of the uncertainty distributions contain the true value. Of course, in practical applications, transfer functions are not computed on lithologic simulations alone. Usually, the lithologic simulations are the first stage of a two-stage simulation procedure designed to incorporate the large-scale geologic features of a region as well as capture the small-scale features of an associated parameter of interest. Both the discrete path and cluster transfer functions reflect the ability of the simulation methods to accurately portray lithologic features. This suggests that perhaps other simulation methods that utilize different summary functions which characterize shape and connectedness (such as Boolean and random set algorithms) might be more appropriate for these transfer functions. For transfer functions such as the discrete average cost function, both categorical simulation algorithms will likely produce accurate response uncertainty distributions. Although uncertainty distributions produced by the truncated Gaussian approach show about the same variability as those produced by categorical sequential indicator simulation, the bias and accuracy measures indicate that the truncated Gaussian approach has a higher average bias. Realizations generated using the truncated Gaussian approach tend to have a larger number of isolated points than those obtained using sequential indicator simulation, and this adversely affects the uncertainty distributions produced by the truncated Gaussian algorithm.

For a particular exhaustive data set, the realizations (taken as an ensemble) produced by the LU decomposition and turning bands algorithms are very similar. The exceptions are the realizations generated using the turning bands algorithim with the Berea exhaustive data set. This particular data set exhibits strong zonal anisotropy that is not parallel to the coordinate axes, and the simplifications made to semivariogram models for use with the turning bands software hindered its ability to portray the banding characteristic of the Berea exhaustive data set. It is surprising that even with several such simplifications, the turning bands algorithm gave fairly accurate results. None of the realizations produced by the turning bands method showed artifacts due to the lines necessary for the one-dimensional simulations.

There are notable visual differences between the realizations produced by the LU decomposition and turning bands methods and those generated using the sequential simulation algorithms. The realizations provided by the sequential methods tend to exhibit more clustering of similar values than realizations produced by the other methods. Figure 3 presents two typical realizations of the Reduced Walker Lake data set; one generated using LU Decomposition, and the other generated using sequential indicator simulation. The clustering is probably an artifact of the kriging on which the sequential approaches are based and is discussed in Hansen (1992) but was not anticipated here since simple kriging was used throughout. Clustering of similar values is particularly pronounced in realizations produced by sequential indicator simulation, and may be due in part to the nugget effect induced in the sequential indicator realizations caused by transformation from the class partitions to the real values. Deutsch and Journel (1992) acknowledge the extra nugget component and suggest artificially reducing the nugget effect of the indicator semivariograms prior to simulation. This suggestion may alleviate part of the problem, but would not affect the clustering problem because it would reduce within class variability and not the extra variability resulting from the discrete partitioning. Figure 3 also illustrates another important difference between the realizations produced by sequential indicator simulation and those obtained by Gaussian-based methods. The use of indicator semivariograms enables realizations of the Reduced Walker Lake data set generated using sequential indicator simulation to capture the ridge of high values characteristic of this data set. Although the Gaussian-based methods capture the large area of high values, the entire ridge is much less clearly defined in these realizations, with no immediate "drop-off" as seen in the exhaustive data set.

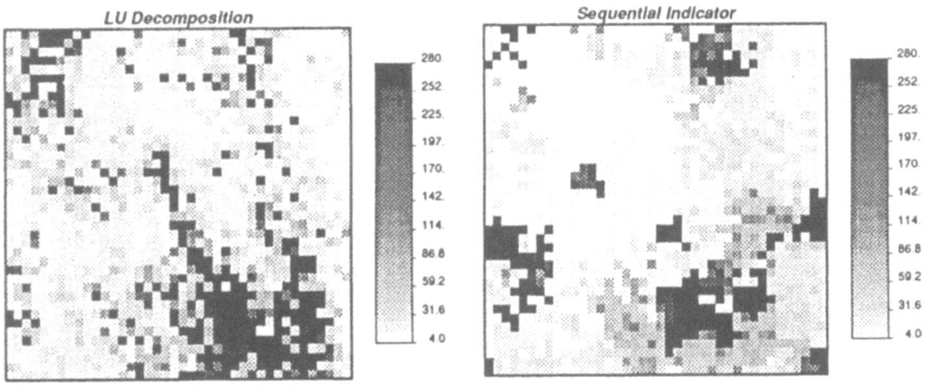

Figure 3. Conditional Realizations of the Reduced Walker Lake Exhaustive Data Set.

The categorical realizations do not usually reflect the shape or position of key features of the categorical exhaustive data sets, even after data-conditioning. Some typical realizations are shown in Figure 4. As mentioned earlier, the conditional realizations obtained using the truncated Gaussian approach have more isolated points than those produced using the categorical sequential indicator approach. This could be due to the details of the implementation of the truncated Gaussian algorithm, such as the choice of an annealing step used in generating the Gaussian realizations or the use of a golden section search algorithm in determining the Gaussian covariance structure, and may not be artifacts of the truncated Gaussian approach in general. At present, there does not appear to be any theory or guidance in the literature as to choices for specific details essential for software implementation of this algorithm or the effect of these choices on the generated realizations.

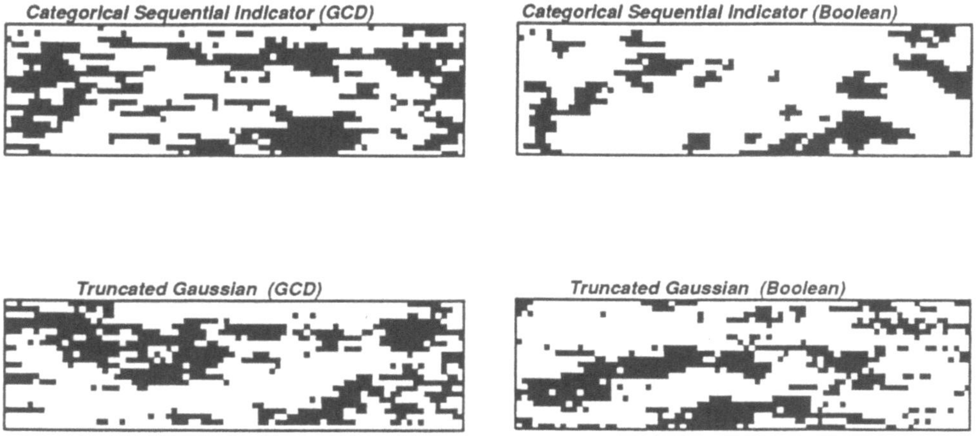

Figure 4. Conditional Realizations of the GCD and Boolean Exhaustive Data Sets.

Different components of the simulation methods appear to affect the shape of the uncertainty distribution of system response. An example is shown in Figure 5, which gives the results of the LU decomposition and sequential Gaussian simulation algorithms for the indicator-covariance specified exhaustive data set using the geometric mean range transfer function. In this situation, the exhaustive data set, transfer function, and summary functions are exactly the same. The sole difference lies with the particular algorithm used. Another example is given in Figure 6, which shows the results of the turning bands and sequential indicator simulation algorithms for the uniform exhaustive data set and the geometric mean average transfer function. It is not clear if the differences between these distributions are due to method of generation (sequential vs. non-sequential), summary function specification (semivariogram vs. indicator semivariograms), or algorithm implementation details. A final example is given in Figure 7, which shows response distributions for the minimum cost transfer function and the Gaussian exhaustive data set obtained using the sequential Gaussian and turning bands algorithms. Here, the spread of the two uncertainty distributions is quite different. These examples, and many others not presented here, show that the uncertainty attributed to a system response prediction can depend on the particular simulation algorithm used and may differ substantially even among methods that are based on the same assumptions and use the same available information.

In general, the effect of data conditioning on the response uncertainty distributions is clear: Conditioning decreases the variability in the system response distribution. In particular, the tails of the distribution are pulled toward the center, the median is generally moved closer to the true value, and most outliers are

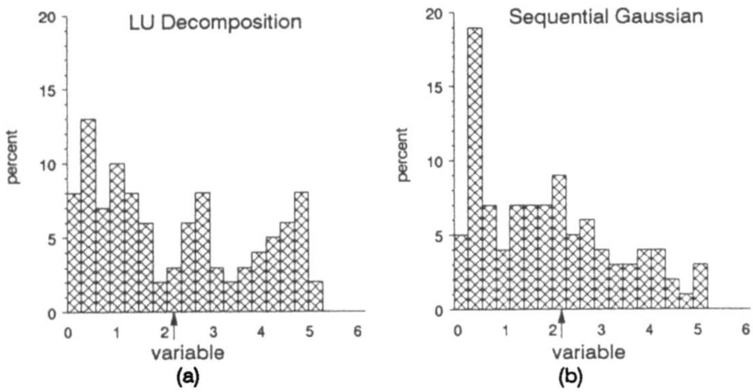

Figure 5. Uncertainty Distributions for the Geometric Mean Range Transfer Function Based on Unconditional Realizations of the Indicator-Covariance Specified Exhaustive Data Set. The arrow denotes the true transfer function value.

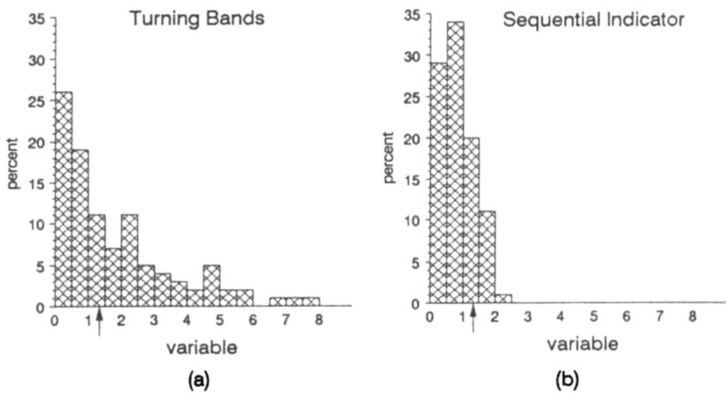

Figure 6. Uncertainty Distributions for the Geometric Mean Average Transfer Function Obtained from Unconditional Realizations of the Uniform Exhaustive Data Set. The arrow denotes the true transfer function value.

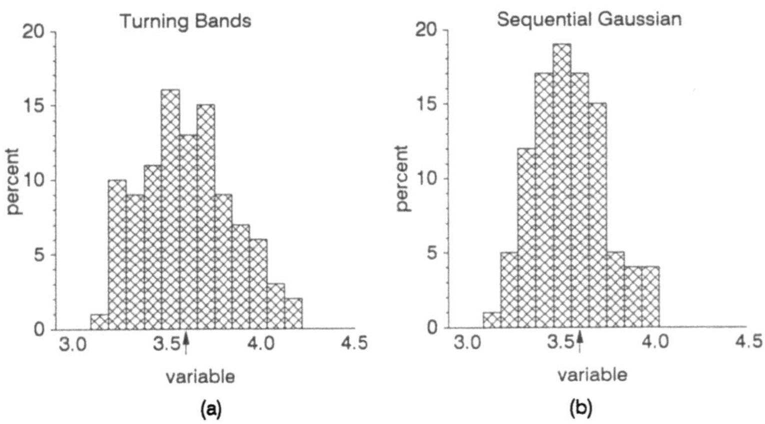

Figure 7. Uncertainty Distributions for the Minimum Cost Transfer Function Obtained from Unconditional Realizations of the Gaussian Exhaustive Data Set. The arrow denotes the true transfer function value.

eliminated (see Figure 8 for a typical example). Conditioning the realizations will also reduce algorithmic effects but not eliminate them. An example is given in Figure 9, which shows uncertainty distributions produced using conditional LU decomposition and sequential indicator simulation algorithms for the indicator-covariance specified exhaustive data set and the geometric mean range transfer function. Here, the two uncertainty distributions are very different even after conditioning on 100 data points. In a few other situations, such as that depicted in Figure 10, it would appear that the increase in precision due to data conditioning was obtained at the expense of bias, since the conditioned distribution no longer contains the true response. This example (and the three other cases observed in this study) support the observations in Hansen (1992), where the tendency was termed "over-conditioning."

All of the simulation algorithms used in this study had the benefit of complete exhaustive information wherever possible. The exceptions were the tail extrapolation choices in the sequential indicator algorithm necessary to extrapolate the conditional distribution estimated at the lowest (highest) threshold to a specified minimum (maximum). As mentioned earlier, the nature of the response distributions produced by the sequential simulation methods can be greatly affected by some parameters that are difficult to determine from the data. Tail extrapolation choices coupled with the choice of specified data minimum and maximum are two such important parameters. In this study, linear extrapolation to the exhaustive data minimum and maximum was used. Results obtained using linear interpolation/extrapolation between exhaustive quantiles (information provided to the other techniques when transforming and back-transforming) produced uncertainty ranges with tighter bounds. Consequently, results based on this interpolation/extrapolation choice were less accurate than those presented here. In addition, a third set of realizations was generated using linear extrapolation to specified minimum and maximum data values that were 10% beyond those given by the exhaustive data. This was done because, in practical applications, exhaustive information will not be available, and a researcher using simulation may decide to let the simulated values fall slightly outside of the data range in order to create the "tails" of the distribution. Based on this third set of realizations, the measures in Table 1 obtained for the Gaussian-based methods did not change much, but some of the measures for sequential indicator were very different. Specifically, most uncertainty distributions had much higher ranges (precision measures were 3.20 (unc.) and 1.43 (cond.)), but were no more accurate than the previous distributions (bias measures were 0.56 (unc.) and 0.30 (cond.)). Clearly, the choice of such parameters is very important, and careful evaluation is required before using realizations generated by any simulation algorithm to assess the uncertainty of a system response.

SUMMARY AND CONCLUSIONS

Several very broad issues are illustrated by the results of this study. The first is the effect of particular simulation algorithms on the resulting uncertainty distribution. Even for simulation algorithms that use the same summary functions, the shape and spread of the resulting uncertainty distributions based on a particular transfer function can be quite different. This lack of consistency implies that different algorithms may produce very different predictions of system response and poses additional problems for applications whose major interest lies in quantifying the uncertainty associated with particular events.

Second, there are some transfer functions (like the discrete path transfer function) for which producing the distribution of system response based on just two summary functions (like the cdf and semivariogram) may not be appropriate. We

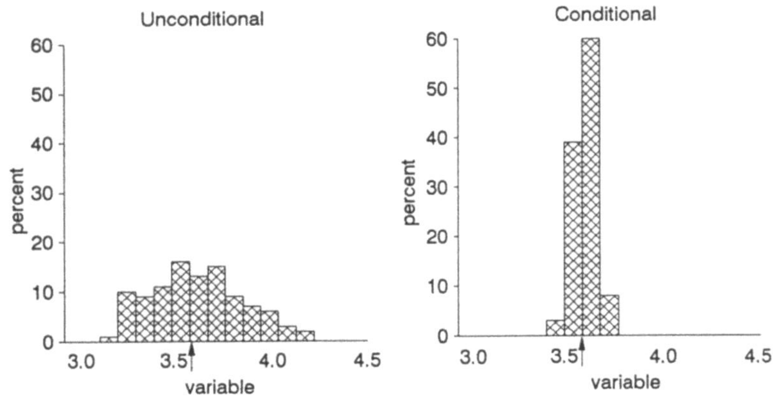

(a) (b)

Figure 8. Uncertainty Distributions for the Minimum Cost Transfer Function Obtained from Realizations of the Gaussian Exhaustive Data Set Generated by the Turning Bands Algorithm. The arrow denotes the true transfer function value.

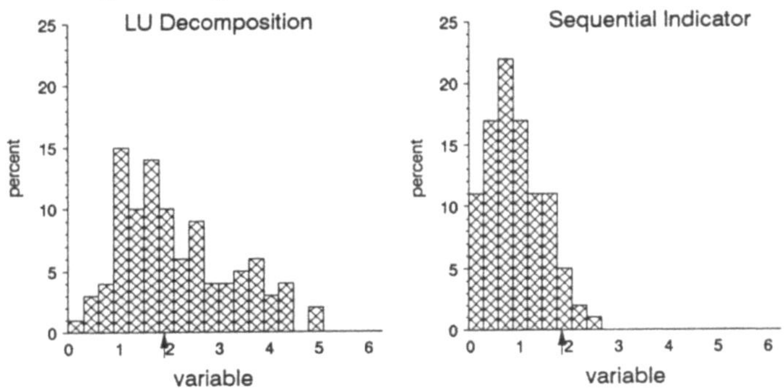

(a) (b)

Figure 9. Uncertainty Distributions for the Geometric Mean Range Transfer Function Obtained from Conditional Realizations of the Indicator-Covariance Specified Exhaustive Data Set. The arrow denotes the true transfer function value.

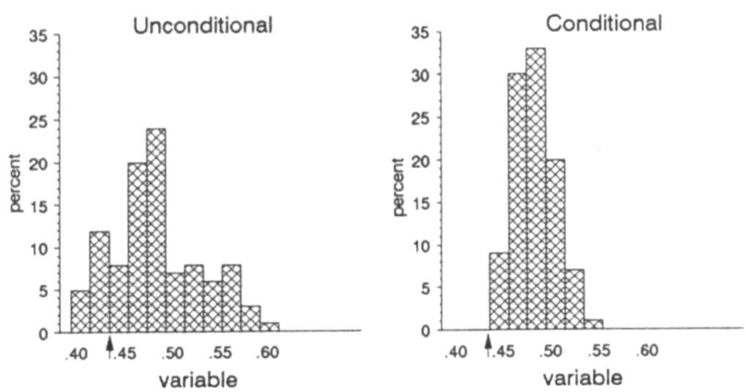

(a) (b)

Figure 10. Uncertainty Distributions for the Minimum Cost Transfer Function Obtained from Realizations of the Berea Sandstone Exhaustive Data Set Generated by Sequential Gaussian Simulation. The arrow denotes the true transfer function value.

observed that for lithologic simulation in which the shape and connectedness of units are important, simulation algorithms that utilize summary functions that can measure these properties might be more appropriate than those considered in this study.

Third, conditioning on sample data will improve the precision associated with the system response distribution and reduce, but not eliminate, algorithmic effects. A few cases of "over-conditioning" were observed but did not appear to be a chronic problem.

Finally, our results indicate that, overall, Gaussian based simulation models can incorporate the essential spatial features of a spatially-varying parameter. Although Gaussian models maximize spatial disorder, and Journel and Deutsch (1993) maintain that they could yield a response uncertainty space that is too narrow, our results do not support this contention. We have found that the Gaussian-based approaches tend to produce uncertainty distributions that are more accurate than those obtained using the sequential indicator algorithm. For imaging spatial uncertainty in a continuous variable, this study suggests that, for the variety of exhaustive-data-set/transfer-function scenarios considered, Gaussian-based approaches, and the sequential Gaussian simulation algorithm in particular, are flexible and accurate methods for stochastic simulation of random fields.

There are many other interesting studies that could have been done within this project, such as considering the effects of summary function estimation from limited data, conditioning sample size and location, and transfer function approximations. For the practicioner, these are real unknowns. To look at these factors over all of the various simulation algorithms would be unnecessarily tedious and redundant. This study provides additional information on the simulation algorithms that appear to work well in a variety of exhaustive data set/transfer function combinations. Studies that look at the effects of other factors not yet investigated can then be designed around these algorithms. In particular, our results suggest that there is still much work to be done in recovering information lost due to discretization of values necessary for the sequential simulation algorithm and in developing simulation methodolgy for categorical variables. In particular, refinements and improvements to the methods illustrated in this paper so that they might more adequately capture continuity and shape, and the study and evaluation of the properties of other categorical simulation approaches such as Boolean and random set algorithms, nearest neighbor approaches, and further development of fractal simulation methods could provide intersting and useful research directions.

ACKNOWLEDGEMENTS

We are grateful to Tony Zimmerman for his assistance with both the implementation of the turning bands algorithm, and the computation of a groundwater travel time transfer function that was not presented in this paper. Comments and suggestions made by Noel Cressie, Bob Easterling, Kathy Hansen, and two reviewers were also very helpful.

REFERENCES

Cressie, N. (1985). "Fitting variogram models by weighted least squares," International Association for Mathematical Geology, 17, 563-586.

Cressie, N. (1991). Statistics for Spatial Data, Wiley, New York.

Davis, M.W. (1987a). "Generating large stochastic simulations via the LU triangular decomposition of the covariance matrix," Mathematical Geology, 19, 91-98.

Davis, M.W. (1987b). "Generating large stochastic simulations–the matrix polynomial approximation method," Mathematical Geology, 19, 99-107.

Desbarats, A.J., and Srivastava, R.M. (1991). "Geostatistical characterization of groundwater flow parameters in a simulated aquifer," Water Resources Research, 27, 687-698.

Deutsch, C.V., and Journel, A. G. (1992a). GSLIB: Geostatistical Software Library and User's Guide, Oxford University Press, New York.

Deutsch, C.V., and Journel, A.G. (1992b). "Annealing techniques applied to the integration of geological and engineering data," Stanford Center for Reservoir Forecasting, Stanford, University.

Dowd, P. A. (1992). "A review of recent developments in geostatistics," Computers and Geosciences, 17, 1481-1500.

Galli, A., Guerillot, D., Ravenne, C., and HERESIM Group, (1990). "Combining geology, geostatistics, and multiphase fluid flow for 3D reservoir studies," in D. Guerillot and O. Guillon (eds.), Proceedings of the 2nd European Conference on the Mathematics of Oil Recovery, Editions Technip, Paris, pp. 11-19.

Giordono, R., Salter, S., and Mohanty, K. (1985). "The effects of permeability variations on flow in porous media," SPE paper 14365, 60th SPE Annual Conference, Las Vegas, NV.

Gomez-Hernandez, J.J. and Srivastava, R.M. (1990). "ISIM3D: An ANSI-C three dimensional multiple indicator conditional simulation program," Computers and Geosciences, 16, 395-440.

Gotway, C.A., Conrad, S.H., Zimmerman, D.A., and McCord, J.T. (1993). "Sequential indicator simulation of lithology with application to vadose zone water flow and transport, Water Resources Research, submitted.

Hansen, K.M. (1992). "The use of sequential indicator simulation to characterize geostatistical uncertainty," Technical Report SAND91-0758, Sandia National Laboratories, Albuquerque, NM.

Isaaks, E.H., and Srivastava, R.M. (1989). Applied Geostatistics, Oxford, New York.

Journel, A.G. (1974). "Geostatistics for conditional simulation of ore bodies," Ecomomic Geology, 69, 673-687.

Journel, A.G. (1983). "Nonparametric estimation of spatial distributions," Journal of the International Association for Mathematical Geology, 15, 445-468.

Journel, A.G. (1989). Fundamentals of Geostatistics in Five Lessons, American Geophysical Union, Washington, D.C.

Journel, A.G., and Alabert, F. (1989). "Non-Gaussian data expansion in the earth sciences," Terra Nova, 1, 123-134.

Journel, A.G., and Deutsch, C.V. (1993). "Entropy and spatial disorder," Mathematical Geology, 25, 329-355.

Kirkpatrick, S., Gelatt, C.D., and Vecchi, M.P. (1983). "Optimization by simulated annealing," Science, 220, 671-680.

Mantoglou, A. and Wilson, J.L. (1982). "The turning bands method for simulation of random fields using line generation by a spectral method," Water Resources Research, 18, 1379-1384.

Matheron, G., Beucher, H., de Fouquet, C. and Galli, A. (1987). "Conditional simulation of the geometry of fluvio-deltaic reservoirs," SPE paper #16753, 62nd Conference of the Society of Petroleum Engineers, Dallas, TX.

Press, W.H., Flannery, B.P., Teukolsky, S.A., and Vetterling, W.T. (1986). Numerical Recipes, Cambridge University Press, Cambridge.

Quimby, W.F. (1986). "Selected topics in spatial analysis: Nonstationary vector kriging, large scale conditional simulation of three dimensional random fields, and hypothesis testing in a correlated random field," PhD dissertation, Department of Statistics, University of Wyoming, Laramie, WY.

Rossi, M., and Posa, D. (1992). "A Non-parametric bivariate entropy estimator for spatial processes," Mathematical Geology, 24, 539-552.

Statistical Analysis System (1990). SAS User's Guide, SAS Institute, Cary, NC.

Zimmerman, D.A., and Wilson, J.L. (1990). Description of and user's manual for TUBA: a computer code for generating two-dimensional random fields via the Turning Bands Method, SEASOFT, Albuquerque, NM.

AN EXTENSION OF THE LU DECOMPOSITION METHOD OF SIMULATION

P.A. DOWD and C. SARAÇ
Department of Mining and Mineral Engineering
University of Leeds
Leeds LS2 9JT, U.K.

For various reasons there has been a number of alternatives proposed to the original turning bands method of geostatistical simulation. Of these, the LU decomposition method is particularly attractive as it is simple to implement, performs conditioning simultaneously with simulation, is not limited to particular forms of covariance functions and automatically handles anisotropies. It is, however, severely restricted by computer memory size requirements. This paper reviews the original LU decomposition algorithm and then proposes an alternative method of decomposition, via ring theory, which significantly extends the size of simulation which can be performed.

INTRODUCTION

Geostatistical simulation was first introduced some 20 years ago in the form of the *turning bands method* (Journel and Huijbregts, 1978). At the time (and since) it was widely expected that simulation would become a major tool in geostatistical applications. However, despite many papers and published programs simulation, with the possible exception of the oil industry, has failed to live up to its promise. This has been largely due to two factors :

- confusion on the part of many end-users (especially in the mining industry) and on the part of some geostatisticians about the meaning and interpretation of simulation results

- inherent problems in the turning bands method

It is largely because of the latter and of the increased interest in the oil industry that the search for alternative methods of simulation has intensified over the past few years.

Several authors have drawn attention to the shortcomings of the turning bands method and the results it sometimes produces. The technique is particularly sensitive to artefacts, such as banding or striping, which are produced by the finite number of lines used in the simulation. The method begins with a simulation of 1-D processes on lines and then projects these simulations onto the required set of 2-D or 3-D co-ordinates where they are summed to give the multidimensional simulation. The method requires an even

M. Armstrong and P. A. Dowd (eds.), Geostatistical Simulations, 23–36.

distribution of lines in the simulation space and, whilst this is a trivial problem in two dimensions, it is much more difficult in three dimensions. The turning bands method uses an icosahedron to generate fifteen lines although attempts have been made, with varying success, to increase this number by defining more lines with an approximately even covering of the three-dimensional space. Additional problems are encountered in simulating anisotropic covariance functions and in the need to perform any conditioning separately from the simulation. The method is also limited by the need to derive a one-dimensional covariance model from the covariance model of the two- or three-dimensional process which means that the usual, space domain version of the turning bands method is restricted to particular forms of the covariance function. The final simulated values, being the sum of a number of 1-D simulated values, are of necessity built up via the gaussian distribution. Some authors maintain that this imparts elements of unwanted gaussian characteristics but no conclusive case has ever been made. Journel and Alabert (1989) draw attention to the maximum entropy property of the bivariate gaussian model showing that it precludes any indicator correlation at extreme thresholds; using an example, they show that such a model conflicts with experimental observation.

Apart from gaussianity and anisotropy and the restrictions on covariance functions, it should be stated that the problems associated with the turning bands method are due to discrete and finite approximations used in the application of the method rather than to faults in the theory. Christakos (1987) shows that the turning bands method is but a particular case of the general space transformation operators thus lending support to the theory as well as to its applicability in the space and frequency domains. A survey of simulation methods is given in Dowd (1992).

Amongst recently proposed alternatives to the turning bands method is Davis' (1987a) LU decomposition method and his related matrix polynomial approximation method (1987b). This method is very simple and overcomes all of the problems of the turning bands method. However, in practice, it is severely restricted by computer memory requirements. First the LU decomposition will be reviewed and than an alternative approach to the decomposition, which reduces the memory problems, will be introduced.

LU DECOMPOSITION METHOD

Consider a set of grid points on which values of random variables are to be simulated. The covariance matrix (of size n x n) of the random variables is denoted by $\mathbf{C} = (C_{ij})$.

The L-U (Lower-Upper) decomposition method is a simple technique based on the L-U triangular decomposition of the matrix of covariances between data locations and simulation grid nodes (Alabert, 1987; Davis, 1987a). Covariance matrices are symmetric and positive-definite and can therefore be decomposed into the product of a lower and an upper triangular matrix :

$$\mathbf{C} = \mathbf{LU} \quad \text{where } \mathbf{L}' = \mathbf{U}$$

Decomposition is by the Cholesky method.

A non-conditional simulation is obtained by multiplying $\mathbf{L}$ by a vector of independent N(0,1) distributed random numbers. Let $\mathbf{w}$ be a vector of independent N(0,1) distributed random numbers and define the vector $\mathbf{y}$ by $\mathbf{y} = \mathbf{Lw}$. The random variable Y corresponding to $\mathbf{y}$ is such that :

$$\begin{aligned} \mathrm{Cov}(Y) &= E[(\mathbf{Lw})(\mathbf{Lw})'] = E(\mathbf{Lww'U}) \\ &= \mathbf{L\,U}\,E(\mathbf{ww'}) \\ &= \mathbf{L\,U\,I} = \mathbf{C} \end{aligned}$$

$\mathbf{y}$ is thus a non-conditional simulation of Z.

A conditional simulation can be obtained as follows.

Partition $\mathbf{C}$ as :

$$C = \begin{pmatrix} C_{DD} & C_{DG} \\ C_{GD} & C_{GG} \end{pmatrix}$$

where $\mathbf{C}_{DD}$: is the covariance matrix between data locations
$\mathbf{C}_{GD}$: is the covariance matrix between grid locations and data locations
$\mathbf{C}_{GG}$: is the covariance matrix between grid locations

This matrix is decomposed as :

$$C = \begin{pmatrix} C_{DD} & C_{DG} \\ C_{GD} & C_{GG} \end{pmatrix} = LU = \begin{pmatrix} L_{DD} & 0 \\ L_{GD} & L_{GG} \end{pmatrix} \begin{pmatrix} U_{DD} & U_{DG} \\ 0 & U_{GG} \end{pmatrix}$$

From this relation, the following equations are obtained:

$$\begin{aligned} \mathbf{C}_{DD} &= \mathbf{L}_{DD}\,\mathbf{U}_{DD} \\ \mathbf{C}_{DG} &= \mathbf{L}_{DD}\,\mathbf{U}_{DG} \\ \mathbf{C}_{GD} &= \mathbf{L}_{GD}\,\mathbf{U}_{DD} \\ \mathbf{C}_{GG} &= \mathbf{L}_{GD}\,\mathbf{U}_{DG} + \mathbf{L}_{GG}\,\mathbf{U}_{GG} \end{aligned}$$

Define $\mathbf{w}$ as the vector $\begin{pmatrix} w_D \\ w_G \end{pmatrix}$, where $\mathbf{w}_G$ is a vector of independent N(0,1) distributed random numbers, and $\mathbf{w}_D$ is the conditioning vector such that

$$\mathbf{L}_{DD}\,\mathbf{w}_D = \mathbf{z}_D$$

where $\mathbf{z}_D$ is the vector of values at data locations (ie, the conditioning data).

Defining $\mathbf{y}$ by $\begin{pmatrix} y_D \\ y_G \end{pmatrix} = \mathbf{Lw}$, gives :

$$\mathbf{y}_D = \mathbf{L}_{DD}\,\mathbf{w}_D = \mathbf{z}_D$$
$$\mathbf{y}_G = \mathbf{L}_{GD}\,\mathbf{w}_D + \mathbf{L}_{GG}\,\mathbf{w}_G$$

the conditional simulation of values at the grid locations is :

$$\mathbf{y}_G = \mathbf{L}_{GD}\,\mathbf{L}_{DD}^{-1}\,\mathbf{z}_D + \mathbf{L}_{GG}\,\mathbf{w}_G$$

where $(\mathbf{L}_{GD}\,\mathbf{L}_{DD}^{-1}\,\mathbf{z}_D)$ is the component that accounts for the conditioning data and $(\mathbf{L}_{GG}\,\mathbf{w}_G)$ the random component that allows any number of simulations to be done.

The advantages of the method are that it is simple to implement, performs conditioning simultaneously with simulation, is not limited to particular forms of covariance functions and automatically handles anisotropies. The drawbacks of the method are the amount of memory required, which increases with the square of the number of grid nodes, and the computing time which increases with the cube of the number of grid nodes. In practice, storage effectively limits the method to two dimensions and less than 1,000 grid nodes.

Some methods of reducing storage limitations

Memory requirements could be reduced by using the technique in conjunction with a moving neighbourhood but this complicates an algorithm for which simplicity is a major advantage; moving neighbourhoods also introduce artefacts into the simulation.

Davis (1987b) introduced a matrix polynomial approximation method as a further development of the LU decomposition approach. This method is based on finding a symmetric matrix $\mathbf{B}$ such that the covariance matrix $\mathbf{C}$ can be factored as :

$$\mathbf{C} = \mathbf{B}\,\mathbf{B} = \mathbf{B}^2$$

then $\mathbf{y} = \mathbf{Bw}$ is also a non-conditional simulation. The problem is in finding the square root of the covariance matrix by a computationally efficient method. The Schur eigenvalue decomposition of the covariance matrix yields :

$$\mathbf{C} = \mathbf{Q}\Lambda\mathbf{Q}'$$

where the columns of $\mathbf{Q}$ are eigenvectors of $\mathbf{C}$ and $\Lambda = \mathrm{diag}(\lambda_1,\lambda_2,\ldots\ldots,\lambda_n)$ and λ_i is the eigenvalue associated with the eigenvector $\mathbf{e}_i$. $\mathbf{B}$ is then defined by :

$$\mathbf{B} = \mathbf{Q}\Lambda^{1/2}\mathbf{Q}'$$

where $\Lambda^{1/2} = \mathrm{diag}(\lambda_1^{1/2},\lambda_2^{1/2},\ldots\ldots.,\lambda_n^{1/2})$. A function of the matrix **C** is defined as :

$$f(C) = \mathbf{QFQ'}$$

where $\mathbf{F} = \mathrm{diag}[f(\lambda_1),f(\lambda_2),\ldots\ldots.,f(\lambda_n)]$ and f is a real, continuous function. The function $f(\lambda_i)$ is then approximated by the minimax polynomial function g(x) on the interval $[\lambda_{min}, \lambda_{max}]$.

Conditioning in the matrix polynomial approximation method is done in the same way as in the turning bands method. Although Davis (1987b) suggests refinements for an efficient implementation of the matrix polynomial method, the major drawback of the technique is still prohibitive computing time and memory for large simulations.

RING DECOMPOSITION

The LU decomposition method requires triangular decompositions of the covariance matrices. When there are many data or when there is a large number of points on which values are to be simulated, the correspondingly large matrices cannot be handled by classical decomposition algorithms. For matrices with dimensions greater than 1000 x 1000, Cholesky decomposition is no longer feasible and a maximum of somewhat less than 1000 simulated values can be generated by the method at any one time.

Ring Decomposition Theory can be applied to reduce significantly this memory-size problem. First an explanation of direct sums of matrices will be given and then the algorithm will be demonstrated.

The relevant aspects of ring decomposition are given in the appendix to this paper. The important result for this application is that if **C** is a matrix ring over the field **R**, then **C** is the direct sum of $\mathbf{B}_i$ where $\mathbf{B}_i$ is the ith row.

As an example suppose that $\mathbf{C} = (C_{ij})$ where (C_{ij}) is an n x n matrix. Let $\mathbf{e}_i = (\mathbf{0},\ldots,\mathbf{0},\mathbf{I},\mathbf{0},\ldots,\mathbf{0})$ **I** is the n x n unit matrix in the ith position. Define $\mathbf{B}_i = \mathbf{e}_i\mathbf{C}$. Then it is easy to see that :

$$\mathbf{C} = \oplus\, \mathbf{B}_i$$
$$\mathbf{e}_i\mathbf{C} = (C_{i1}, C_{i2}, C_{i3}, \ldots\ldots\ldots., C_{in})$$

since the vector $(C_{i1},\ldots\ldots.,C_{in})$ can be identified with the $\mathbf{B}_i$, it follows that

$$\begin{aligned}\mathbf{C} &= \mathbf{e}_1\mathbf{C} \oplus \mathbf{e}_2\mathbf{C} \oplus \ldots\ldots\ldots \oplus \mathbf{e}_n\mathbf{C} \\ &= \mathbf{B}_1 \oplus \mathbf{B}_2 \oplus \ldots\ldots\ldots \oplus \mathbf{B}_n\end{aligned}$$

Next, the aim is to decompose $\mathbf{B}_i$ as the product of the lower and upper triangular matrices :

$$\begin{pmatrix} 0 & 0 & \dots & 0 \\ \cdot & \cdot & & \cdot \\ \cdot & \cdot & & \cdot \\ C_{i1} & C_{i2} & \dots & C_{in} \\ \cdot & \cdot & & \cdot \\ \cdot & \cdot & & \cdot \\ 0 & 0 & \dots & 0 \end{pmatrix} = \begin{pmatrix} L_{11} & 0 & \dots & 0 \\ L_{21} & L_{22} & \dots & 0 \\ \cdot & \cdot & & \cdot \\ \cdot & \cdot & & \cdot \\ \cdot & \cdot & & \cdot \\ \cdot & \cdot & & \cdot \\ L_{n1} & L_{n2} & \dots & L_{nn} \end{pmatrix} \begin{pmatrix} U_{11} & \cdot & \dots & U_{1n} \\ 0 & U_{22} & \dots & U_{2n} \\ \cdot & \cdot & & \\ \cdot & \cdot & & \cdot \\ \cdot & \cdot & & \cdot \\ \cdot & \cdot & & \cdot \\ 0 & 0 & \dots & U_{nn} \end{pmatrix}$$

Note that the LU decomposition of a matrix **C** is not unique, and the following decomposition can be used :

$$\begin{aligned} L_{11}U_{11} &= C_{11} && \text{(by Cholesky decomposition)} \\ L_{i1} &= C_{i1} && i \geq 2 \\ L_{ij} &= C_{i,j-1} - C_{ij} && 2 \leq j \leq i-1 \\ L_{ii}U_{ii} &= C_{i,i-1} - C_{ii} && i \geq 2 \ \text{(by Cholesky)} \\ L_{ij} &= 0 && j > i \end{aligned}$$

Therefore, $\mathbf{B}_i = \mathbf{L}_i\mathbf{U}_i$ (with $\mathbf{L}_i' = \mathbf{U}_i$ as in the standard LU decomposition) and $\mathbf{L}_i$ is obtained by the process described above.

The method can be explained by means of a simple example. Consider a 500 (nx) x 3 (ny) simulation grid, ie a total of nx x ny = 1500 grid points. Although this is an unrealistic array it will serve as an example on which the relevant equations can be developed by hand.

$$\begin{pmatrix} C_{DD} & C_{DG} \\ C_{GD} & C_{GG} \end{pmatrix}$$

Here $\mathbf{C}_{GG}$ is a 1500 x 1500 covariance matrix which is too large for classical decomposition methods to be used to calculate the LU matrices. The problem can, however, be solved by defining sub-matrices corresponding to each fixed value of ny (or, in general, any combination of fixed values of one or two dimensions) :

$$C_{GG} = \begin{pmatrix} C_1 & C_2 & C_3 \\ C_4 & C_5 & C_6 \\ C_7 & C_8 & C_9 \end{pmatrix}$$

Each sub-matrix has dimensions 500 x 500 (nx x nx) and there are 3 x 3 (ny x ny) sub-matrices.

Applying the block decomposition technique of Ring Theory to C_{GG} gives:

$$C_{GG} = e_1 C_{GG} \oplus e_2 C_{GG} \oplus e_3 C_{GG}$$

where e is the unit idempotent matrix, here $e_1 = (I,0,0)$, $e_2 = (0,I,0)$ and $e_3 = (0,0,I)$.

$$C_{GG} = \begin{pmatrix} C_1 & C_2 & C_3 \\ 0 & 0 & 0 \\ 0 & 0 & 0 \end{pmatrix} + \begin{pmatrix} 0 & 0 & 0 \\ C_4 & C_5 & C_6 \\ 0 & 0 & 0 \end{pmatrix} + \begin{pmatrix} 0 & 0 & 0 \\ 0 & 0 & 0 \\ C_7 & C_8 & C_9 \end{pmatrix}$$

Solution for ny = 1

$$\begin{pmatrix} C_1 & C_2 & C_3 \\ 0 & 0 & 0 \\ 0 & 0 & 0 \end{pmatrix} = \begin{pmatrix} L_1 & 0 & 0 \\ A_1 & L_2 & 0 \\ A_2 & A_3 & L_3 \end{pmatrix} \begin{pmatrix} U_1 & B_1 & B_2 \\ 0 & U_2 & B_3 \\ 0 & 0 & U_3 \end{pmatrix}$$

From this realisation, the following equations are obtained:

$$\begin{array}{llll} C_1 = L_1U_1 & A_1U_1 = 0 & A_2U_1 & = 0 \\ C_2 = L_1B_1 & A_1B_1 + L_2U_2 = 0 & A_2B_1 + A_3U_2 & = 0 \\ C_3 = L_1B_2 & A_1B_2 + L_2B_3 = 0 & A_2B_2 + A_3B_3 + L_3U_3 & = 0 \end{array}$$

$$\begin{pmatrix} C_1 & C_2 & C_3 \\ 0 & 0 & 0 \\ 0 & 0 & 0 \end{pmatrix} = \begin{pmatrix} L_1 & 0 & 0 \\ 0 & 0 & 0 \\ 0 & 0 & 0 \end{pmatrix} \begin{pmatrix} U_1 & L_1^{-1}C_2 & L_1^{-1}C_3 \\ 0 & I & I \\ 0 & 0 & I \end{pmatrix}$$

Solution for ny = 2

$$\begin{pmatrix} 0 & 0 & 0 \\ C_4 & C_5 & C_6 \\ 0 & 0 & 0 \end{pmatrix} = \begin{pmatrix} L_1 & 0 & 0 \\ A_1 & L_2 & 0 \\ A_2 & A_3 & L_3 \end{pmatrix} \begin{pmatrix} U_1 & B_1 & B_2 \\ 0 & U_2 & B_3 \\ 0 & 0 & U_3 \end{pmatrix}$$

From this realisation, the following equations are obtained:

$$\begin{array}{llll} L_1U_1 = 0 & C_4 = A_1U_1 & A_2U_1 & = 0 \\ L_1B_1 = 0 & C_5 = A_1B_1 + L_2U_2 & A_2B_1 + A_3U_2 & = 0 \\ L_1B_2 = 0 & C_6 = A_1B_2 + L_2B_3 & A_2B_2 + A_3B_3 + L_3U_3 & = 0 \end{array}$$

$$\begin{pmatrix} 0 & 0 & 0 \\ C_4 & C_5 & C_6 \\ 0 & 0 & 0 \end{pmatrix} = \begin{pmatrix} 0 & 0 & 0 \\ C_4 & L_{C_5-C_4} & 0 \\ 0 & 0 & 0 \end{pmatrix} \begin{pmatrix} I & I & I \\ 0 & U_{C_5-C_4} & L^{-1}_{C_5-C_4}(C_6-C_4) \\ 0 & 0 & 0 \end{pmatrix}$$

Solution for ny = 3

$$\begin{pmatrix} 0 & 0 & 0 \\ 0 & 0 & 0 \\ C_7 & C_8 & C_9 \end{pmatrix} = \begin{pmatrix} L_1 & 0 & 0 \\ A_1 & L_2 & 0 \\ A_2 & A_3 & L_3 \end{pmatrix} \begin{pmatrix} U_1 & B_1 & B_2 \\ 0 & U_2 & B_3 \\ 0 & 0 & U_3 \end{pmatrix}$$

From this realisation, the following equations are obtained:

$$\begin{array}{llllll} L_1U_1 & = 0 & A_1U_1 & = 0 & C_7 = A_2U_1 \\ L_1B_1 & = 0 & A_1B_1 + L_2U_2 & = 0 & C_8 = A_2B_1 + A_3U_2 \\ L_1B_2 & = 0 & A_1B_2 + L_2B_3 & = 0 & C_9 = A_2B_2 + A_3B_3 + L_3U_3 \end{array}$$

$$\begin{pmatrix} 0 & 0 & 0 \\ 0 & 0 & 0 \\ C_7 & C_8 & C_9 \end{pmatrix} = \begin{pmatrix} 0 & 0 & 0 \\ 0 & 0 & 0 \\ C_7 & C_8-C_7 & L_{C_9-C_8} \end{pmatrix} \begin{pmatrix} I & I & I \\ 0 & I & I \\ 0 & 0 & U_{C_9-C_8} \end{pmatrix}$$

where $L_{C_k-C_l}$ and $U_{C_k-C_l}$ are, respectively, the lower and upper matrices of C_k-C_l obtained by Cholesky decomposition.

From the equation for the conditional simulation of values at the grid points :

$$y_G = L_{GD}\, L_{DD}^{-1}\, z_D + L_{GG}\, w_G$$

The L_{GG} matrices have been obtained for each ny. When calculating the L_{GG} matrices, L_{GD} and L_{DD}^{-1} matrices can be obtained for each ny. In this manner a simulated value is obtained at each grid point.

Note that the sub-matrices can be established for any subdivision of a two- or three-dimensional simulation volume. The subdivision chosen will depend on the specific two- or three-dimensional array of grid points on which values are to be simulated and does not necessarily have to coincide with the increments of one of the dimensions (eg, ny = 1,2,3).

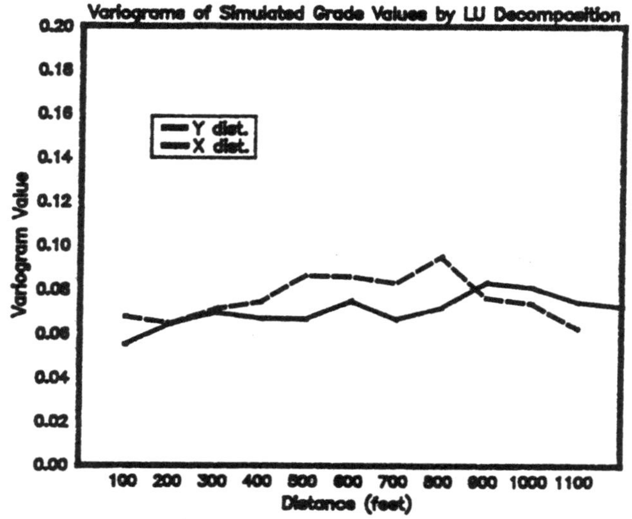

Figure 1

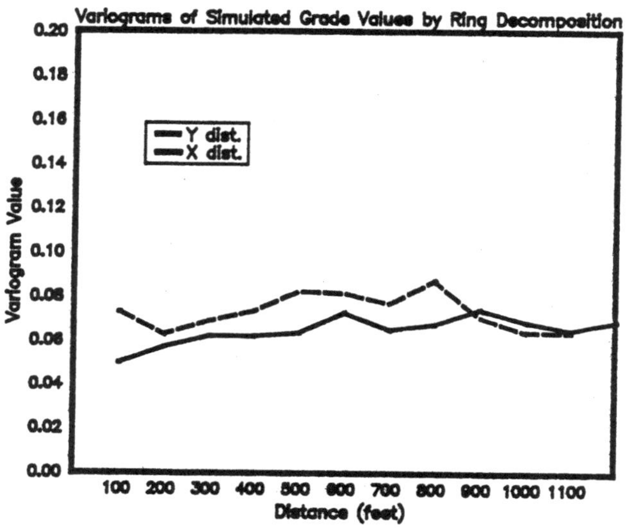

Figure 2

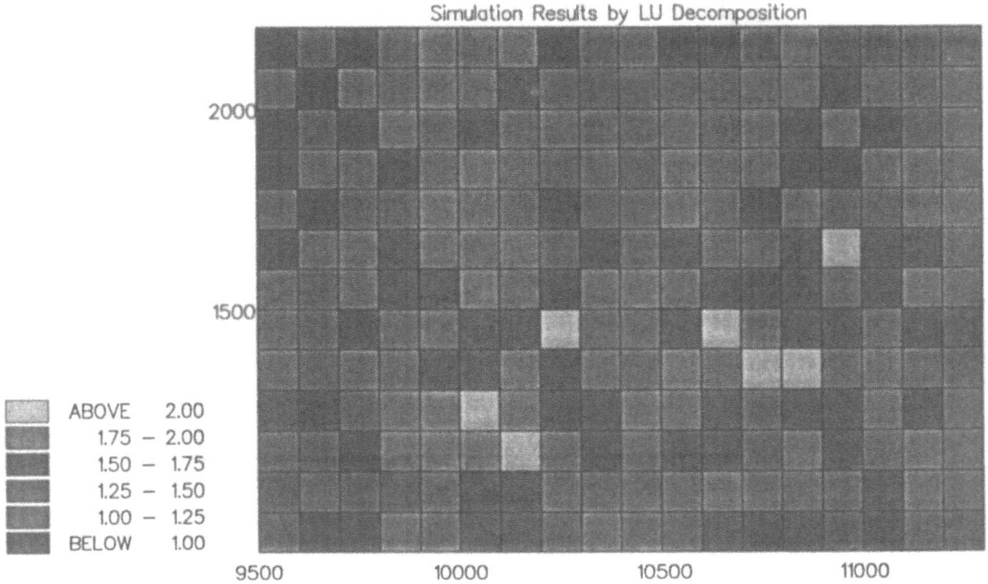

Figure 3

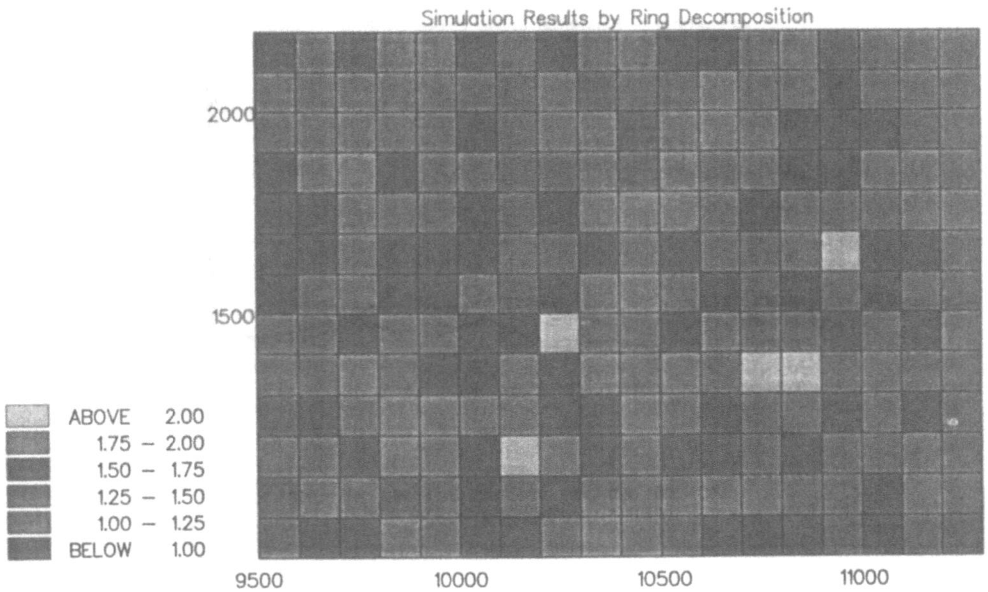

Figure 4

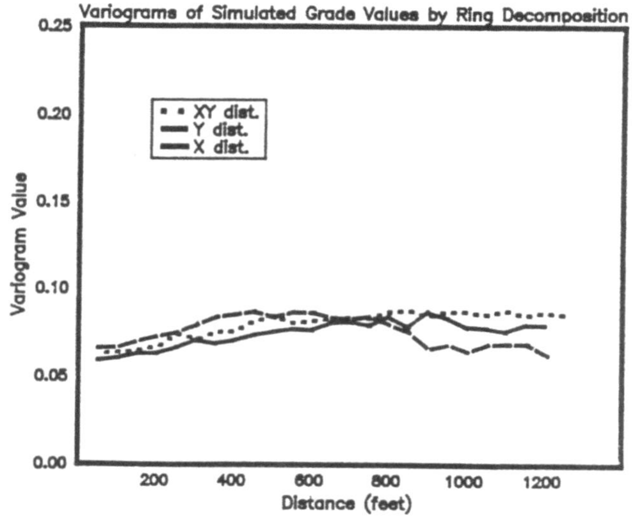

Figure 5

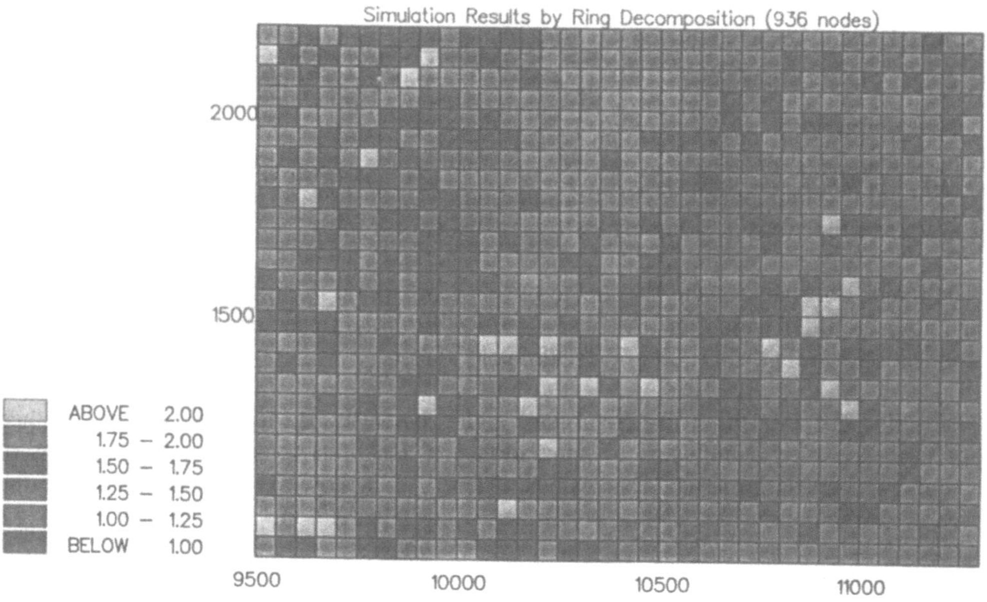

Figure 6

EXAMPLE

This example is taken from a published data set given in David (1988) page 156*ff*. A total of 77 intersections with a copper vein are recorded over an area of 1800ft x 1300ft. The data values are approximately normal with mean 1.40%Cu and standard deviation 0.28%Cu. The experimental variograms are rather noisy; an isotropic variogram model was fitted with parameters $C_0 = 0.035(\%)^2$, $C = 0.040(\%)^2$ and range 600ft. David uses the turning bands method to perform a conditional simulation of 972 values on a 50ft x 50ft grid. The authors found that storage requirements made it impossible to perform this simulation via the standard form of LU decomposition. For the purpose of providing a comparison, simulations on a 100ft x 100ft grid were performed using both the standard LU decomposition and the ring decomposition. These results are summarised in figures 1 - 4. Next the ring decomposition method was used for the full simulation on the 50ft x 50ft grid. These results are summarised in figures 5 and 6.

Note that both the methods (figures 1 - 4) yield satisfactory simulations. The 50ft x 50ft simulation by ring decomposition also produces satisfactory results which are in accordance with the results from the turning bands method as reported by David. Computing times, on a 8-10 mips rated machine, for the 100ft x 100ft grid are 25 cpu secs for Cholesky decomposition and 36 cpu secs for ring decomposition; the full 50ft x 50ft simulation takes 320 cpu secs. The differences in computing times are typical although the times are reduced significantly on a higher rated machine.

CONCLUSIONS

The advantages of the LU method are that it is simple to implement, performs conditioning simultaneously with simulation, is not limited to particular forms of covariance functions and automatically handles anisotropies.

The major drawback of the LU method using Cholesky decomposition is the amount of memory required which, at least in its general form as presented, effectively limits its application to less than 1000 grid locations.

Using ring decomposition, the covariance matrix between grid locations can be obtained for each specified subdivision of simulation grid locations thus providing a means of simulating very large numbers of values. There is still, however, an upper limit on the size of simulations which can be performed using this method.

REFERENCES

Alabert, F. (1987) The practice of fast conditional simulations through the LU decomposition of the covariance matrix. *Journal of Mathematical Geology*, vol. 19, no 5, pp 369 - 386.

Anderson, F. W. and Fuller, K. R. (1974) Rings and categories of modules. Springer-

Verlag New York Inc, 340 p.

Christakos, G. (1987) Stochastic simulation of spatially correlated geo-processes. *Jour. Math. Geology*, vol. 19, no. 8, p 807-831.

David, M. (1988) Handbook of applied geostatistical ore reserve estimation. Developments in Geomathematics No. 6, Elsevier, Netherlands, 216p.

Davis, M. (1987a). Production of conditional simulations via the LU triangular decomposition of the covariance matrix. *Mathematical Geology*, vol 19, no. 2, pp 91-98.

Davis, M. (1987b) Generating large stochastic simulations - the matrix polynomial approximation method. *Mathematical Geology*, volume 19, no. 2, pp 99-108.

Dowd, P.A. (1992) A review of recent developments in geostatistics. *Computers and Geosciences* vol. 17, No. 10, pp 1481-1500.

Journel, A.G. and Alabert, F. (1989) Non- gaussian data expansion in the earth sciences. *Terra Nova*, vol 1 p 123-134.

Journel, A.G. and Huijbregts, C. (1978) Mining Geostatistics. Academic Press, New York, 600p.

Kasch, F. (1982) Modules and Rings. Translation and editing by D. A. R. Wallace, Academic Press, 372 p.

APPENDIX : RING DECOMPOSITION

An *R-module* is a generalisation of a vector space where the scalars are allowed to come from a ring rather than a field. An additive subgroup of an R-module is a *sub-module*. The following is taken from Kasch (1982) and rests on the observation that vector spaces, and consequently, R-modules, are direct sums of the same simple sub-module.

M is called the direct sum of the set $\{B_i \mid i \in I\}$ of sub-modules $B_i \hookrightarrow M$, in symbols:

$$M = \bigoplus_{i \in I} B_i \leftrightarrow \begin{cases} 1.)\ M = \sum_{i \in I} B_i \\ 2.)\ \forall j \in I\ [B_j \cap \sum_{\substack{i \in I \\ i \neq j}} B_i = 0] \end{cases}$$

where $\sum_{i \in I} B_i = \{\sum_{finite} b_i \mid b_i \in B_i\}$

$M = \bigoplus_{i \in I} B_i$ is also said to be a direct decomposition of M into the sum of the sub-modules $\{B_i \mid i \in I\}$.

Here it is assumed for all $i \in I$, $0 \in B_i$.

In the case of a finite index set, say $I = \{1, \ldots\ldots, n\}$, M is also written as $M = B_1 \oplus \ldots\ldots \oplus B_n$ (Kasch, 1982)

THEOREM : Let $\{B_i \mid i \in I\}$ be a set of sub-modules $B_i \in M$ and let $M = \Sigma_i B_i$. Then (2) of the previous direct sum definition is equivalent to :

for every $x \in M$ the representation

$$x = \sum_{i \in I'} b_i \quad \text{with} \quad b_i \in B_i,\ I' \subset I, \quad I' \ \text{finite}$$

is unique in the following sense :

$$\text{If} \quad x = \sum_{i \in I'} b_i = \sum_{i \in I'} c_i \quad \text{with} \quad b_i,\ c_i \in B_i$$

then it follows that

$$\forall i \in I', \quad b_i = c_i$$

(The proof of the theorem can be found on page 151, Anderson and Fuller, 1974).

COROLLARY : If **M** is a matrix ring over the field **R**, then **M** is the direct sum of $\mathbf{B}_i$ where $\mathbf{B}_i$ is the ith row.

DISCUSSION FOLLOWING SESSION NO 1

Chairman: Jean–Paul Chilès

Papers: Gotway & Rutherford ; Dowd

After the end of the second paper, the chairman thanked the speakers for very interesting presentations and asked if there were any questions on the second paper (by Dowd).

Gomez: Is the LU decomposition of the total covariance matrix the same as the sum of the LU decompositions of the components of the direct sum?

Dowd No. In the ring decomposition:

$$C = L\,U$$
$$= C_1 \oplus C_2 \oplus \ldots \oplus C_n$$
$$= L_1\,U_1 \oplus L_2U_2 \oplus \ldots . \oplus L_n\,U_n$$

But, if **L** is defined as:

$$L = L_1 + L_2 + \ldots.. + L_n$$

and **U** is defined as:

$$U = U_1 + U_2 + \ldots. + U_n$$

then

$$C \neq L\,U$$

I think that your question amounts to asking whether ring decomposition yields the same simulation as LU decomposition. The answer is no. However there is no unique LU decomposition of a covariance matrix and each decomposition would yield an equally valid, but different, simulation. Similarly ring and LU decom– positions yield different but equally valid simulations. There is, however, a very high correlation (in excess of 95%) between simulated values obtained by LU decomposition and those obtained from the corresponding ring decomposition.

M. Armstrong and P. A. Dowd (eds.), Geostatistical Simulations, 37–42.

Note, though, that the LU decomposition of a direct sum of matrices is the same as the direct sum of the LU decompositions of each summand.

Gomez: In your example you worked level by level.

Dowd: The subdivision could have been done in other ways. One should choose the most efficient one (i.e. the one that would result in the fewest submatrices). For example, in the case of the 500 x 3 x 10 one, it would have been better to have split it into vertical slices (i.e. for each NZ, then changing NY).

Freulon: In the examples presented, the variogram model is close to a pure nugget effect. Has the method been tested on more structured cases?

Dowd: We selected David's data set simply because it had already been published. The tests were carried out initially on very structured data sets. Four sets have been used to date.

Ravenscroft: Is it true that all the conditioning data are used all the time in the D matrix? Is that then a limitation of the method?

Dowd: We have assumed all the way through the number of conditioning data is not large; however, if it was, the method could still be applied but it would have to be applied to the D matrix as well, which would make it even more complex.

Ravenscroft: Would it help in taking subsets of data in the neighbourhood around the point to be simulated?

Dowd: Yes. That suggestion was made by Davis for the original LU decomposition but it makes the method cumbersome. One of the method's great advantages is its simplicity. The way that we have set up the program at present, the end user does not see how the ring decomposition is made. It is transparent for him. Implementing neighbourhoods would change this.

de Chambure: Is it possible to generate many simulations using this method?

Dowd: Yes it is, as it is with the original LU decomposition. This is not a new method, it is just an extension of an existing one.

de Chambure: What about nongaussian data?

Dowd: Yes you do have to make a transformation.

de Chambure: Is there any overlap between the rings?

Dowd: The technique is very flexible. There is no restriction in the way you subdivide the matrix. I do not believe that this a problem. One merely has to redefine the split up.

The chairman then threw the discussion open to questions on both papers.

Daly: In her paper Carol Gotway talks about overconditioning. I ran into this problem myself. You have got to calculate the variogram on the appropriate data set. Out of laziness I did not recalculate the histogram and variogram on the data. When I went back and did this, I did not find the problem arose any longer. So Carol, could you till us whether you recalculated the histogram and the variogram on each data set?

Gotway: No we did not, essentially because we do not want to introduce any bias or subjectivity by having to model the empirical variogram from the finite sample sets.

Gomez to Daly: Do you mean that by recomputing the variograms from the data you got the distribution to fit the real values?

Daly: Yes because essentially the histogram and variogram from 100 data is not the same as that for 20 000 data. There is quite a lot more variability. Repeating each simulation with the variogram from that data set leads to a larger spread. It is also better centered.

Gotway to Daly:In our approach we did not choose just 1 set of conditioning points, because they might not have been representative. We chose 100 such sets. Does you solution still apply, given the number of sets used?

Daly: I'm not sure. Clearly some of the 100 sets will be more typical; others will not and will give quite an atypical variogram.

Omre: We are discussing a very important aspect of the problem. First we have reality, which we can evaluate. Then we have a stochastic model. There is a considerable gap between these two. The largest problem is to come up with a model that is really representative of reality. But our problem today is how to simulate the model. I think that when making their presentations the authors should try to focus on how we can reproduce realizations from our model because though we are all interested in modelling reality, the uncertainty due to modelling will be much larger than the errors due to our simulation algorithms (and that is as it should be). So I think that it is very important to specify our model clearly and then focus on our simulation algorithms.

Armstrong: Carol Gotway has considered 4 functions for assessing her simulations. These are applicable to environmental questions and hydrology, but are they also appro– priate in other fields such as mining? For example, is minimum path length the most interesting criterion for mining? What should we use to characterize whether a simulation is, or is not, reproducing what we want?

Omre: It is a loss function that you are asking for and this it highly dependent on the use. Water flow is very different from mining.

Chilès: It is not possible to give a unique answer. It depends upon the application. In particular, it depends on much more than just the histogram and the covariance. A specified algorithm produces a realization of a random function with a given multivariate distribution. So it can be more suitable in some cases.

Coming back to Carol Gotway's paper, Chilès thanked her for a very thought provoking presentation but noted that it was not possible to draw general conclusions from it.

Rivoirard: It seems to me that people often choose a simulation algorithm without testing to see whether the model that it is supposed to reproduce, actually suits the data. For example, the sequential indicator model is not the same as the diffusion models. While checking the model makes things more complicated, it is nevertheless very important.

Omre: I agree with Jacques Rivoirard and re–iterate the importance of properly specifying the model.

Chilès, as chairman, agrees on the importance of a suitable model.

Ravenscroft: Firstly I would like to congratulate Carol Gotway and Brian Rutherford on their paper. It is a very brave thing to publish these kinds of results. I think that a number of us have been tempted to do it in the past. But just a word to add to previous discussion. We have talked about specifying the model underlying the algorithm, but the other stage is to specify how the algorithm is implemented. Anyone who has used turning bands, or sequential methods, or decomposition methods, will know that these are very sensitive to the way it is done. It concerns me that ten years in the future your paper will be used to show that one method is better another. Categoric statements of that sort should not be allowed to go without some sort of rider about the parameters or the software used in this particular case.

Gotway: I agree whole heartedly. In the first case, we tried to improve on the case studies that already exist. There are several that have been done on 1 exhaustive data set and 1 transfer function, from which general conclusions have been drawn. Clearly you cannot draw completely general conclusions from 1 case study with 1 transfer function. You can do a little bit better by consider several exhaustive data sets and several transfer functions. However the way you interpret the results does depend on the application envisaged.

There has to be a certain subjectivity as to how to implement the algorithms, and also in the computer code used. Brian and I wrote some of the software ourselves and modified existing codes. There were many guidelines available for implementing the software that we used. The authors of software will be the most knowledgeable about implementation aspects but this should not preclude others from using the technique.

Your points are well taken but I would nevertheless refrain from dismissing the study by saying that no general conclusions can be drawn. You can draw some conclusions. The differences in the algorithms really do exist.

Ravenscroft: I am certainly not dismissing what you have done. I am very impressed. I am just saying that one must remember that these things are not entirely general.

Chilès: I have a question on the comparison between LU decomposition and the sequential gaussian method. You say that the two methods are very different. In my opinion, both methods are equivalent because they produce multivariate normal fields except that the LU method is global and uses a unique neighbourhood whereas sequential methods need a moving neighbourhood. So I am surprised that there is such a big difference, and would like to know whether other people have experienced the same thing. Have you any idea why there is so much difference?

Gotway: My explanation is the same as yours. Firstly the LU decomposition is global whereas the other is local. Then the sequential methods are sequential, to begin with, which the other is not. These are the only differences. We back transformed them both in exactly the same way.

Ravenscroft: I would like to come back to the question of overconditioning. Clearly it is very worrying for practitioners to think that conditional simulations are susceptible to the quantity of data you give them; the more you give them the less accurate they become. Could we have some more discussion about this and about what we are going to do about it?

Note from the Editors: At the end of session N°1, Colin Daly agreed to give some more detailed information in writing, on overconditioning. Here it is:

I agree with Carol Gotway that "overconditioning" does exist. However I do not believe that the name "overconditioning" is the correct one. It seems to me (but I could be wrong) that the problem is essentially due to the fact that the gaussian model is not appropriate for the data at hand.

We should consider the two cases: with & without conditioning data.

(1) Without conditioning data Suppose we have at our disposal an observation which depends on many points of our realisation. As an example, suppose our random function is binary taking values of 0 and 1, and that we have some indirect method of measuring the size of the smallest connected cluster of 1's. Furthermore let us assume that the true random function is Boolean of fixed grain size. The smallest connected cluster is thus at least the size of the grain. Now suppose that we try to model this random function without knowing that it is Boolean and we mistakenly use a truncated gaussian model. Typically the smallest connected cluster for the truncated gaussian will be very small. So that the histogram of responses from many trials will not contain the observed minimum connected cluster size. We thus have a mismatch without any conditioning data!

(2) With conditioning data This effect gets worse as the variance drops. The histogram gets narrower. Moreover, when using a gaussian model the conditional expectation of a nonlinear function is biased if the data are not multigaussian. Since the conditional expectation estimate may be written in terms of Hermite polynomials as

$$(Y)^{EC} = \eta_0 + \sum \eta_i \; H_n(Y^*)$$

it can be shown in the usual way that

$$E[f(Y)^{EC}] = \eta_0 \Leftrightarrow \sum \lambda_i Y_i = Y^* \text{ are Gaussian } \Leftrightarrow \{Y_i\} \text{ multigaussian}$$

Hence, as the conditional expectation gives the mean value used in subsequent simulations, the results are even further biased. Note that disjunctive kriging estimates do not suffer from this problem of bias, thereby ensuring that at least we can do unbiased estimation in practice So to summarize, I think that the effect that is called "overconditioning", is due to the choice of an inappropriate model and is further exacerbated by the conditioning effect of data.

STOCHASTIC SIMULATION AND CONDITIONING BY ANNEALING IN RESERVOIR DESCRIPTION

BJØRN KÅRE HEGSTAD*, HENNING OMRE*,
HÅKON TJELMELAND* and KELLY TYLER†

* The Norwegian Institute of Technology
Division of Mathematical Science
N-7034 Trondheim
Norway

† Statoil a.s.
Po box 300
N-4001 Stavanger
Norway

1 Introduction

Simulation of realizations of high dimensional probability distributions is often complicated. In the continuous case under Gaussian assumptions, several well understood algorithms are available. In the general case the picture seems to be fairly confusing.

There exist several more or less general iterative simulation algorithms, such as the Metropolis algorithm, Gibbs sampler, Hastings algorithm, Simulated Annealing (SA) and Threshold Accepting (TA). To be able to select a suitable algorithm for a given model, it is important to understand the properties and limitations of the different algorithms. One should understand both which algorithms are able to sample from the model specified and the efficiency of the algorithms for this model. Several of the points in this paper are more elaborated in Hegstad (1993).

The paper is based on the results on SA in Geman and Geman (1984), but utilizes it for simulation purposes. This have been suggested by others in the petroleum sector, see Farmer (1989), Deutsch (1992) and Ouenes et al. (1992), and they focus on the optimization as a way of fulfilling constraints. We emphasize that it is the knowledge of the probability over the states fulfilling the constraints, which is the important result in Geman and Geman (1984). Moreover, the paper presents a type of stochastic models which easily can be conditioned to complex constraints.

M. Armstrong and P. A. Dowd (eds.), Geostatistical Simulations, 43–55.

2 Simulation by Hastings algorithms

There exist several algorithms for generating realizations from a probability distribution π with state space Ω. The Gibbs sampler algorithm, Geman and Geman (1984) and the Metropolis algorithm, Metropolis et al. (1953), are familiar examples. Both these algorithms are special cases of a class of simulation algorithms described by Hastings (1970). The idea is to construct a Markov chain with Ω as state space and π as limiting distribution.
Hastings prescribes the following approach for constructing such a chain: Let $P = \{p_{ij}\}$ be the transition matrix of the homogeneous, aperiodic and irreducible Markov chain with state space $\Omega = \{0, 1, \ldots, N\}$. Then if X_k denotes the state occupied by the process at time k, one has:

$$Pr\{X_{k+1} = j | X_k = i\} = p_{ij}$$

Let an arbitrary matrix element p_{ij} have the form:

$$p_{ij} = q_{ij} \cdot \alpha_{ij} \qquad \text{for} \qquad i \neq j$$

$$p_{ii} = 1 - \sum_{j \neq i} p_{ij}$$

where $\mathbf{Q} = \{q_{ij}\}$ is the transition matrix of an arbitrary irreducible Markov chain with the same state space Ω. Let α_{ij} be given by

$$\alpha_{ij} = \frac{s_{ij}}{1 + \frac{\pi_i}{\pi_j}\frac{q_{ij}}{q_{ji}}}$$

where $\pi = (\pi_0, \pi_1, \ldots, \pi_N)$ is a probability distribution with $\pi_i > 0$ for all i. The parameter s_{ij} may be chosen freely under the constraints

$$s_{ij} = s_{ji} \tag{1}$$

$$0 < s_{ij} < 1 + \frac{\pi_i}{\pi_j}\frac{q_{ij}}{q_{ji}} \tag{2}$$

The latter ensures that $0 < \alpha_{ij} < 1$, such that α_{ij} can be considered as a probability. This defines an iterative algorithm which in each iteration draws a potentially new state with probabilities specified by the Q-matrix, and then accepts this potentially new state with the probability α_{ij}.

From basic Markov theory, see Taylor and Karlin (1984), it follows that π will be the limiting distribution of such a Markov chain, since

$$\pi P = \pi$$

defines the limiting distribution. A sufficient condition for this is

$$\pi_i p_{ij} = \pi_j p_{ji}, \qquad \forall i,j \in \Omega$$

By using Hastings construction above one gets:

$$\begin{aligned} \pi_i p_{ij} &= \pi_i q_{ij} \alpha_{ij} = \pi_i q_{ij} \frac{s_{ij}}{1 + \frac{\pi_i}{\pi_j} \frac{q_{ij}}{q_{ji}}} \\ &= \pi_i q_{ij} \frac{s_{ij}}{\frac{\pi_i}{\pi_j} \frac{q_{ij}}{q_{ji}} \left(\frac{\pi_j}{\pi_i} \frac{q_{ji}}{q_{ij}} + 1 \right)} \\ &= \pi_j q_{ji} \frac{s_{ji}}{1 + \frac{\pi_j}{\pi_i} \frac{q_{ji}}{q_{ij}}} \\ &= \pi_j q_{ji} \alpha_{ji} = \pi_j p_{ji} \end{aligned}$$

Hence one may use the Hastings approach to generate realizations from an arbitrary distribution π.

Both the Metropolis algorithm and the Gibbs sampler algorithm are special cases of the Hastings algorithm with s_{ij} chosen as large as possible:

$$s_{ij} = \min \left\{ 1 + \frac{\pi_i}{\pi_j} \frac{q_{ij}}{q_{ji}}, 1 + \frac{\pi_j}{\pi_i} \frac{q_{ji}}{q_{ij}} \right\} \tag{3}$$

such that

$$\alpha_{ij} = \min \left\{ 1, \frac{\pi_j}{\pi_i} \frac{q_{ji}}{q_{ij}} \right\} \tag{4}$$

In the Metropolis algorithm the Q-matrix is chosen to be symmetric, i.e. $q_{ij} = q_{ji}$ for all $i, j \in \Omega$. The acceptance probability α_{ij} will then become

$$\alpha_{ij} = \min \left\{ 1, \frac{\pi_j}{\pi_i} \right\}$$

In the Gibbs sampler algorithm one choses the q_{ij}'s to be conditional distributions such that $\alpha_{ij} = 1$, hence all new states j are being kept. These conditional distributions are obvious in some models and impossible to find in other. For more details see Geman and Geman (1984) and Hegstad (1993).

3 Simulation by Simulated Annealing

One often wants to constrain the realizations being generated. The realization may be required to reproduce a variogram, well observations, net gross or size distributions. These constraints can be expressed in a so called objective function $V(i)$, which is a measure of the distance between the characteristics of the realization, and the desired constraints. In other words, one wants to generate realizations from a state space $\Omega^* \subseteq \Omega$ where the constraints are fulfilled. The state space Ω^*, is the set of states $i \in \Omega$ where the objective function $V(i) = 0$, or more general, the set of states $i \in \Omega$ where the objective function has its global minima. The latter is significant if it is impossible to fulfill all the constraints simultaneously.

Minimizing $V(i)$ is equivalent to maximizing the function

$$\pi_i = c_0 e^{-V(i)}, \qquad c_0 > 0$$

due to the monotonic behaviour of the exponential function. In the following this fact is important.

For many combinatorial problems, $V(i)$ has one or a few global minima, so Ω^* contains just one or a few states. In reservoir description, one may have many degrees of freedom in the model, and only a limited number of constraints. Hence Ω^* is likely to be a relative large space. The state space Ω^*, will normally be more like a space with islands of prohibited states due to constraints, than islands of permitted states as in combinatorial problems.

Introduce a parameter T, often termed the temperature, and define the probability distribution

$$\pi_i^T = c(T) e^{-\frac{V(i)}{T}} \qquad i \in \Omega \tag{5}$$

with $c(T)$ being a normalizing factor. It is now possible to use a Hastings algorithm for given $T > 0$, to construct a Markov chain with π^T as limiting distribution.

From expression (5) it is seen that for states $i \in \Omega$ where $V(i)$ does not have a global minimum, the function $\pi_i^T \to 0$ as $T \to 0$. The normalizing factor $c(T)$, ensures that not the entire function converges to zero. Define π^* as the uniform distribution on Ω^*. A theorem given by Geman and Geman (1984) describes how slow the parameter T should be decreased, the annealing schedule, to ensure that the corresponding Markov chain has π^* as limiting distribution. The proof is done in the case when Gibbs sampler is used iteratively as T decreases. This is the Gibbs-Hastings Simulated Annealing. The theorem in Geman and Geman (1984) can be put more formally this way: If T_k is any decreasing sequence of temperatures for which

$T_k \to 0$ as $k \to \infty$

$T_k \geq \frac{d}{\ln(k+1)}$

where k is the iteration number, and $d = \max_i V(i) - \min_i V(i)$. Then for any initial state $i \in \Omega$ and every $j \in \Omega$

$$\lim_{k\to\infty} Pr\{X_k = j | X_1 = i\} = \pi^*_j$$

where π^* is the uniform distribution on Ω^*.

The property that after having applied Hastings SA the resulting solution is a realization from a uniform probability distribution over Ω^*, is the important one in stochastic simulation. The fact that the constraints are favored is of course important, but that can be obtained in many ways. Controlling the probability distribution over the allowed state space Ω^*, is much more complicated, and a prerequisite in stochastic modeling. Slow convergence rate is often a problem when Hastings SA is applied to generate such realizations. Suggestions have been made in order to make the algorithm more efficient. This is to some extent discussed in section 6.

4 Stochastic Model in Reservoir Description

The reservoir description of a specific reservoir, is based on a large variety of information spanning from understanding of sedimentary processes, to measurements on core plugs in the reservoir under study. It seems convenient to distinguish two classes of information: phenomenological information and reservoir specific observations. The former includes knowledge of the sedimentary processes creating the reservoir, experience from comparable geological environments and reservoirs, and results from fluid flow studies in schematic reservoirs of similar type. The latter includes all types of observations made on the reservoir under study, for example seismic data, core measurements, well logs and production tests.

The phenomenological information is used for establishing the stochastic model. This information help deciding if mosaic, event or continuous models are to be used, see Omre (1992), and which variables that are worth including. Moreover the phenomenological information is used for assigning prior probability distributions to the model parameters in a Bayesian setting, see Omre et al. (1991). This constitutes the phenomenological model which represents the geological interpretation of the reservoir under study, although without conditioning on the reservoir specific observations. This could for shale units in a sand matrix be a marked point model, see Stoyan et al. (1987). Each marked point, $\underline{u}_s = ((x,y,z)_s, l_s, w_s, h_s, \varphi_s)$, represents an arbitrary three dimensional horizontal shale box with location, (x,y,z); length, l; width, w; height, h; and direction, φ. Define the stochastic variable $\underline{U} = (\underline{U}_1, \ldots, \underline{U}_n)$ and the corresponding realization $\underline{u} = (\underline{u}_1, \ldots, \underline{u}_n)$. Then the phenomenological model being a simultaneous pdf of n shale units may be parameterized by

$$\mathcal{V}_{\underline{U}}(\underline{u}) = c_1 \cdot \exp\left\{-\sum_{r=1}^{n} a(\underline{u}_r) - \sum_{r=1}^{n}\sum_{s=r+1}^{n} b(\underline{u}_r, \underline{u}_s)\right\} \tag{6}$$

with c_1 being a normalizing constant, $a(\cdot)$ being the uni-marked-point component, and $b(\cdot,\cdot)$ being the bi-marked-point component. Higher order components could be included in the parameterization. For other geological interpretations, other model types or parameterization could be preferable.

The reservoir specific observations provides constraints on the realizations being feasible. Observations in wells or other characteristics shall be reproduced in all legitimate realizations. For shale units in a sand matrix this could be the location of shales in a well, iw; sand/gross ratio, sg; etc. Let the corresponding characteristics computed from the realization of marked points be $\widehat{iw}(\underline{u})$ and $\widehat{sg}(\underline{u})$. In the previous section it was shown that for the example above

$$\begin{aligned}\pi^*_{\underline{U}}(\underline{u}) &= \lim_{T\to 0} \pi^T_{\underline{U}}(\underline{u}) \\ &= \lim_{T\to 0} c_2(T)\cdot\exp\left\{-\frac{\left(iw-\widehat{iw}(\underline{u})\right)^2 + (sg-\widehat{sg}(\underline{u}))^2}{T}\right\}\end{aligned}$$

will be a pdf with uniform probability in the domains where the constraints are fulfilled and zero elsewhere. In this case the objective function described in section 3 is given by

$$V(\underline{u}) = \left(iw-\widehat{iw}(\underline{u})\right)^2 + (sg-\widehat{sg}(\underline{u}))^2$$

For other types of constraints, the objective function has of course to look different. Note that in this section only the formulation of the stochastic model is treated. Hence considerations like convergence in simulation algorithms are not made.

So the conditional distribution for the stochastic variable $\underline{U}$ modeling the reservoir is

$$f_{\underline{U}|C}(\underline{u}|c) = c_3 \cdot \mathcal{V}_{\underline{U}}(\underline{u}) \cdot \pi^*_{\underline{U}}(\underline{u}) \tag{7}$$

with $C = c$ indicating the reservoir specific constraints actually imposed, i.e. the objective function $V(\underline{u}) = 0$. One may also present the model as

$$\begin{aligned}f_{\underline{U}|C}(\underline{u}|c) &= \lim_{T\to 0} f^T_{\underline{U}|C}(\underline{u}|c) \\ &= \lim_{T\to 0}\left[c_3(T)\cdot\mathcal{V}_{\underline{U}}(\underline{u})\cdot\pi^T_{\underline{U}}(\underline{u})\right] \\ &= \mathcal{V}_{\underline{U}}(\underline{u})\cdot\lim_{T\to 0}\left[c_3(T)\cdot\pi^T_{\underline{U}}(\underline{u})\right]\end{aligned}$$

For a given value of $T > 0$, realizations from the pdf $f^T_{\underline{U}|C}(\underline{u}|c)$ may be generated by a Hastings algorithm as defined in section 2. In Figure 1 the annealing applied on π^T and the multiplication with ν is illustrated in a simple one-dimensional case. The result is a function non-zero only on the state space Ω^*.

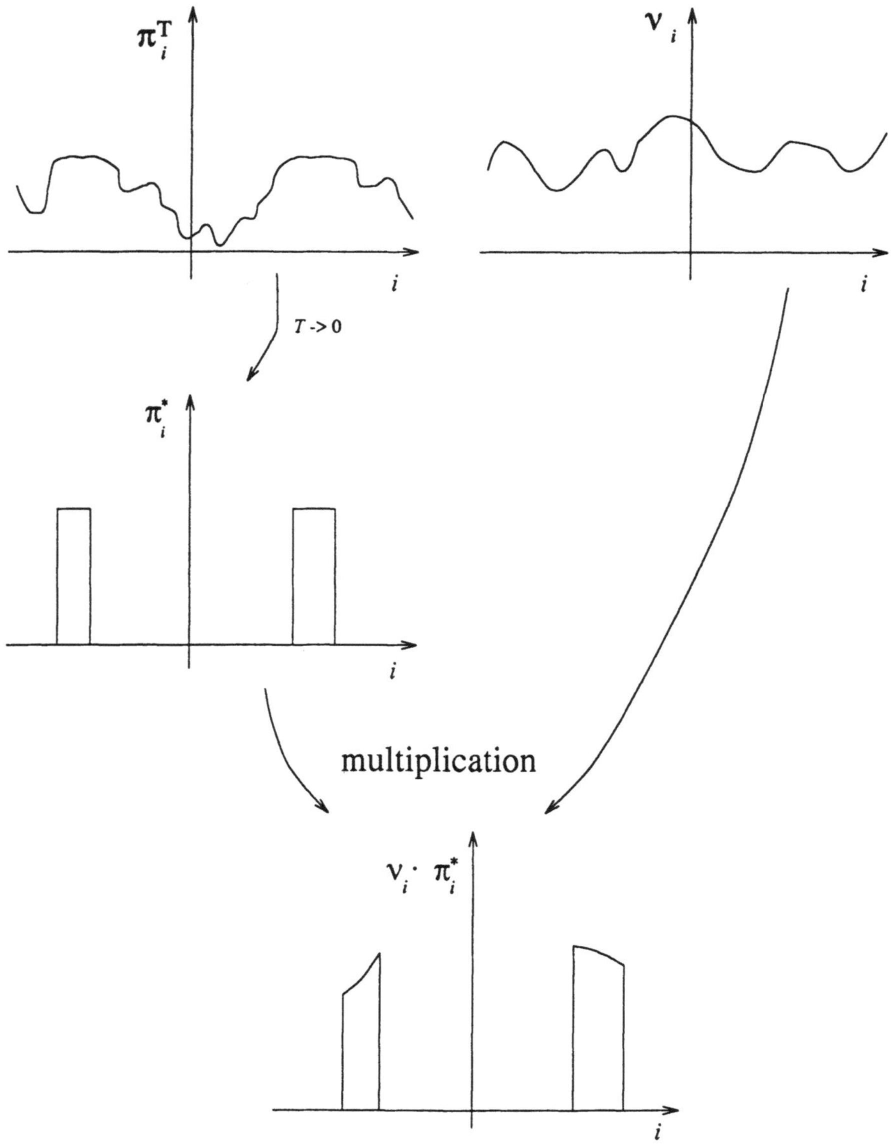

Figure 1: For a simple one-dimensional case the probability distribution π^T is annealed and then multiplied by the probability ν. The result is a function non-zero only on the state space Ω^*.

5 Simulation from the Stochastic Model

Realizations from the pdf

$$f^T_{\underline{U}|C}(\underline{u}|c) = \tag{8}$$

$$c(T) \cdot \exp\left\{ -\sum_{r=1}^{n} a(\underline{u}_r) - \sum_{r=1}^{n} \sum_{s=r+1}^{n} b(\underline{u}_r, \underline{u}_s) \right\} \cdot \exp\left\{ -\frac{\left(iw - \widehat{iw}(\underline{u})\right)^2 + (sg - \widehat{sg}(\underline{u}))^2}{T} \right\}$$

which is expression (7) written out, shall be generated. According to the Hastings algorithm, see section 2, an arbitrary transition matrix $Q = \{q_{ij}\}$ can be chosen without changing the limiting distribution π. The choice of Q will have considerable impact on the convergence rate of the algorithm, however. Good choices will be close to the corresponding conditional probability distributions. Usually Q is defined to have transition probabilities positive for states j close to the present state i, and zero elsewhere. In the Metropolis algorithm and the Gibbs sampler algorithm, the q_{ij}'s are usually non-zero only for transitions along one dimension at a time. In image analysis this entail changing the value in one pixel at a time only. In the example above this would correspond to changing one arbitrary mark, e.g. length l, of one arbitrary point at the time.

The stochastic model in the example above is based on marked points and it is intuitively reasonable to define Q based on this. A natural choice is to have q_{ij} non-zero only for changes in one marked point at the time. This is along the lines of the Ripley-Kelly simulation algorithm for marked point models, see Ripley (1987).

Define the states

$$\begin{aligned} i &= (\underline{u}_1, \ldots, \underline{u}_k, \ldots, \underline{u}_n) \\ j &= (\underline{u}_1, \ldots, \underline{u}'_k, \ldots, \underline{u}_n) \end{aligned}$$

and the transition probability

$$q_{ij} = c \cdot \exp\{-a(\underline{u}'_k)\} \tag{9}$$

See also Hegstad (1993). This is a natural choice given by the model. In this case it is the multivariate distribution for a shale in the reservoir with respect to position and marks. The q_{ij} is generating a new position and a new set of marks for shale number k. The improved efficiency will only be realized if the pdf q_{ij} is such that realizations from it can be generated directly. This can be done if a sequential simulation algorithm based on conditional pdf's for the marks can be used, see Omre et al. (1993). Using this q_{ij} the $a(\cdot)$-terms vanishes and the corresponding acceptance probability of j with respect to i becomes

$$\alpha_{ij} = \frac{s_{ij}}{1 + \frac{f^T_{\underline{U}|C}(i|c)}{f^T_{\underline{U}|C}(j|c)} \exp\{-a(\underline{u}'_k) + a(\underline{u}_k)\}}$$

$$= \frac{s_{ij}}{1 + \exp\left\{\sum_{r=1}^{n}\left[b(\underline{u}_k', \underline{u}_r) - b(\underline{u}_k, \underline{u}_r)\right]\right\} \cdot \exp\left\{-\frac{d_{ij}\left(iw - \widehat{iw}(\underline{u})\right) + d_{ij}\left(sg - \widehat{sg}(\underline{u})\right)}{T}\right\}}$$

with $d_{ij}(\cdot)$ being the difference in squared deviation of the corresponding constraint by replacing i by j. Note in particular that the factor $c(T)$ being difficult to assess, is not involved in the relation.

According to Hastings algorithm the factor s_{ij} can be chosen freely under the constraints (1) and (2). Choosing s_{ij} as large as possible as in (3), i.e.

$$s_{ij} = \min\left\{1 + \frac{f_{\underline{U}|C}^T(i|c)}{f_{\underline{U}|C}^T(j|c)} \exp\left[-a(\underline{u}_k') + a(\underline{u}_k)\right], 1 + \frac{f_{\underline{U}|C}^T(j|c)}{f_{\underline{U}|C}^T(i|c)} \exp\left[-a(\underline{u}_k) + a(\underline{u}_k')\right]\right\}$$

will give the following acceptance probability for j with respect to i:

$$\alpha_{ij} = \min\left\{1, \frac{f_{\underline{U}|C}^T(j|c)}{f_{\underline{U}|C}^T(i|c)} \exp\left[-a(\underline{u}_k') + a(\underline{u}_k)\right]\right\}$$

This is not the Metropolis algorithm since q_{ij} in (9) in general is not symmetric. Nor is it the Gibbs sampler algorithm. The choice of q_{ij} in (9) makes the acceptance probability α_{ij} generally not equal 1. In this particular case the algorithm will correspond to the Ripley-Kelly algorithm, see Ripley (1987).

The steps in the algorithm is as follows:

- Initiation - construct n marked points with positive cdf, e.g. uniformly random.

Iterate several times:

- Select an arbitrary marked point $\underline{u}_k$.
- Generate a new marked point $\underline{u}_k'$ from the pdf $c \cdot \exp\{-a(\underline{u}_i)\}$
- Replace $\underline{u}_k$ by $\underline{u}_k'$ with probability α_{ij} .

Stop iterating.

By operating on marked points and not on individual variables, one will make the algorithm more efficient. The marked points in q_{ij} can be drawn from the uni-marked-point cdf directly without involving that in the acceptance step.

Note that if an $\underline{u}_k$ ensuring the conditioning of reservoir specific observations is selected, the probability of generating a new $\underline{u}_k'$ fulfilling the constraints is negligible.

Hence if T is small the acceptance probability for $\underline{u}'_k$ will be very small and hence the convergence slow. This problem can be solved by importance sampling in domains where the probability of generating a $\underline{u}'_k$ fulfilling the constraints is larger. The importance sampling will be highly dependent on the constraints actually used.

A theorem in Winkler (1990) implicitely states that if an appropriate annealing schedule is followed, the limiting distribution resulting from any Hastings based SA, is equal the limiting distribution resulting from Gibbs-Hastings SA. The latter is the one treated in Geman and Geman (1984). So by using an appropriate annealing schedule, realizations from

$$f_{\underline{U}|C}(\underline{u}|c) = \lim_{T\to 0} f^T_{\underline{U}|C}(\underline{u}|c)$$

can be generated.

6 Threshold Accepting

In the recent years Threshold Accepting (TA) has been suggested as an alternative to the Metropolis-Hastings SA. TA can be considered as a simplification of the latter. The acceptance probability α_{ij} for the Metropolis-Hastings SA in the case (5), is given by

$$\alpha_{ij} = \begin{cases} 1 & , V(j) \leq V(i) \\ e^{-\frac{V(j)-V(i)}{T}} & , V(j) > V(i) \end{cases} \tag{10}$$

while for TA it is given by

$$\alpha_{ij} = \begin{cases} 1 & , V(j) \leq V(i) + T \\ 0 & \text{else} \end{cases} \tag{11}$$

This simplifies the simulation algorithm. The random number generation and calculation of the exponential function, are avoided in the accepting part of the iteration. These simplifications make the calculations at each iteration faster to execute.

TA has been used in combinatorial problems as a optimization algorithm. Dueck and Scheuer (1990) report that in a given traveling salesman problem, TA gives better results than Metropolis-Hastings SA in fewer cpu-seconds.

The montonic decline algorithm for optimization, see Luenberger (1989), is a special case of TA with $T = 0$. It will though lack many of the properties making TA useful in some optimization problems.

A problem is that TA does not have a theoretical basis as does Hastings SA. It is for example readily verified that TA at fixed T is not a Hastings algorithm. TA can though in some settings be considered as an approximation to a class of optimization algorithms described by Hajek (1988). Hajek proves a theorem which ensures that one global optimum is reached with probability one using an algorithm from this class. The theorem does not however say anything about with what probability each global optimum is reached in the multiple optima case, as Geman and Geman's theorem does for Gibbs-Hastings SA.

It may in some cases, be possible to apply TA to generate realizations from Ω^*. But there are no theorems stating from which distribution on Ω^* TA eventually is sampling, i.e. what the limiting distribution of the Markov chain defined by the TA algorithm is. This is an important issue in stochastic reservoir modeling.

One could make the conjecture that TA samples from the uniform distribution on Ω^* in some cases. This could be a reasonable conjecture when Ω^* is connected and got a simple structure. But even then, TA is not a good alternative to Hastings SA in the model described in section 4. Here it was significant to sample from a given distribution $f_{\underline{U}|C}(\underline{u}|c)$ on Ω^* which in general is not the uniform distribution but a combined pdf. This is as far as one knows, not possible using TA, since TA is not a Hastings algorithm.

7 Closing remarks

The class of Hastings algorithms provides a way for generating realizations from general stochastic models. There are several parameters in the algorithm and these can be tuned to the model at hand to obtain better efficiency. The celebrated Simulated Annealing may be used for conditioning on reservoir specific observations, but numerous other alternatives exists. The crucial aspect is, however, to control the probability distribution in the state space fulfilling the conditioning. To obtain this, Hastings based Simulated Annealing algorithms have to be used. These are the theoretically based conclusions. In practice, however, the ideal annealing schedule cannot be followed, hence perfect convergence will not be obtained. Hence the picture is more confusing and careful considerations must be made in each individual case.

References

Deutsch, C.V. (1992): "Annealing Techniques Applied to Reservoir Modeling and the Integration of Geological and Engineering (Well Test) Data " A dissertation submitted to the department of applied earth sciences and the committee on graduate studies of Stanford University in partial filfillment of the degree of Doctor of Philosphy.

Dueck, G. and Scheuer, T. (1990) "Threshold Accepting: A General Purpose Optimization Algorithm Appearing Superior to Simulated Annealing" *Journal of Computational Physics, p.161-175*

Farmer, C.V. (1989) "The Mathematical Generation of Reservoir Geology" Paper presented at the 1989 joint IMA/SPE conference on the Mathematics of Oil Recovery, Cambridge.

Geman, S. and Geman, D. (1984) "Stochastic Relaxation, Gibbs Distribution, and the Bayesian Restoration of Images" *IEEE Transaction on Pattern Analysis and Machine Intelligence, vol. PAMI-6, p.721-741*

Hajek, B. (1988) "Cooling schedules for optimal annealing" *Mathematics of operations research, **13**, No 2*

Hastings, W.K. (1970) "Monte Carlo sampling methods using Markov chains and their applications" *Biometrika, **57**,1, p.97-109*

Hegstad, B.K. (1993) "Stokastisk Modellering og Annealing i Reservoarbeskrivelse" (in Norwegian) Thesis for the degree Master of Science The Norwegian Institute of Technology

Luenberger, D.C. (1989) Linear and Nonlinear Programming, Second Edition, Addison-Wesley Publishing Company, Reading

Metropolis, N., Rosenbluth, A., Teller, A. and Teller, E. (1953) "Equation of state calculation by fast computing machines" *Journal of Chemical Physics, **21**, p.1087-1092*

Omre, H (1992) "Stochastic Models for Reservoir Characterization" in Skjæveland, S.M. and Kleppe, J. (eds.) SPOR Monograph, Recent Advances in Improved Oil Recovery Methods for North Sea Sandstone Reservoirs. Norwegian Petroleum Directorate, Stavanger

Omre, H., Sølna, K. and Tjelmeland, H. (1993) "Simulation of Random Functions on Large Lattices" in Amilcar Soares (ed.) Geostatistics Troia '92, Kluwer Academic Publishers, Dordrecht.

Omre, H., Tjelmeland, H., Qi, Y. and Hindeaker, L. (1991) "Assessment of Uncertainty in the Production Characteristics of a Sand Stone Reservoir" in Bill Linville (ed.), Reservoir characterization III, PennWell Books, Tulsa, Oklahoma.

Ouenes, A., Bahralolom, I., Gutjahr, A. and Lee, R. (1992) "Conditioning Permeability Fields by Simulated Annealing" in M.A. Christie, F.V. Da Silva, C.L. Farmer, O. Guillon, Z.E. Heinemann, P. Lemonnier, J.M.M. Regtien and E. van Spronsen (eds.) ECMOR III, 3rd European Conference on the Mathematics of Oil Recovery, Delft university Press.

Ripley, B.D. (1987) Stochastic Simulation, John Wiley & Sons Inc., New York

Stoyan, D., Kendall, W.S. and Mecke, J. (1987) Stochastic Geometry and its Applications, John Wiley & Son Inc., Chichester.

Taylor, H.M., Karlin, S. (1984) An introduction to stochastic modeling Academic Press, Orlando.

Tjelmeland, H. and Holden, L. (1993) "Semi-Markov Random Fields" in Amilcar Soares (ed.) Geostatistics Troia '92, Kluwer Academic Publishers, Dordrecht.

Winkler, G. (1990) "An Ergodic L^2-theorem for Simulated Annealing in Bayesian Image Reconstruction " *Journal of Applied Probability, 28, p. 779-791.*

Conditional simulation of a Gaussian random vector with non linear and/or noisy observations

Xavier Freulon
20, rue Saint Jean
75 017 Paris
France
Tel : (33-1) 42 63 75 91

ABSTRACT: *We consider the problem of simulating a multigaussian random vector observed through non linear and/or corrupted sensors. To sample the conditional distribution, linear methods such as* ***conditional kriging*** *or* ***sequential Gaussian simulation*** *can be applied only to the case of linear observations. But more complex observations can be integrated into stochastic imaging via the* ***Gibbs sampler****: the conditional distribution is sampled as the stationary distribution of a Markov chain. This is illustrated on three examples: inequality constraints, Poisson noise and 1D convolution.*

Key words: Bayesian analysis, Markov chain, Gibbs sampler, inequality constraints, Poisson noise, deconvolution.

1 Introduction

In geostatistics, conditional simulations are used to estimate by Monte-Carlo methods complicated non linear functions of more than one point which depend explicitly on a multivariate distribution. A classical simulation[12] is a numerical model which honors the experimental data with a given marginal distribution and a fixed covariance function. It relies on a specific model (the variable under study is interpreted as a particular realization of a one-to-one univariate transform of a Gaussian field) and accurate measurements (at every experimental point the value of the variable is exactly known).

This article presents methods to condition a Gaussian model with a wider class of observations — non linear and/or noisy observations. The aim is to incorporate into the simulation all the available data, i.e. to sample a conditional distribution which

M. Armstrong and P. A. Dowd (eds.), Geostatistical Simulations, 57–71.

is "*data charged*" and internally consistent. The calculation of this conditional law can be achieved within the Bayesian analysis framework.

The paper is divided in three sections: first we present the class of observations we considered, then we state some remarks on the well known linear case. Finally the Gibbs sampler is used to simulate the general case.

2 The model and the observations

In the following, X denotes the vector to be simulated and Z the vector of the observations. Their respective dimensions are n and p, both finite. The model is based on two assumptions:
Assumption 1: The vector X is a centered Gaussian vector with a specified covariance matrix, Σ.

$$X \sim \mathcal{N}(0, \Sigma)$$

Assumption 2: We know the conditional distribution of the observed records z given any particular value of the random vector X.

$$Z|x \sim dF(z|x)$$

With this model, a conditional simulation of X given a value of the observations is a realization of the conditional distribution of X given z. In Bayesian terms[4], the prior probability density is the Gaussian probability density function, $g(x)$, the likelihood of the vector X for observed records z is the conditional probability density function of Z given x whereas a conditional simulation of X given z is a sample of the posterior distribution.

The first assumption derives from a model widely used in geostatistics: the studied variable is interpreted as a realization of the univariate transform of a stationary Gaussian field[1], $(\phi(X_u))_{u \in D}$. But a numerical model is a finite collection of samples (e.g. a square grid), so we consider the Gaussian vector

$$X = (X_{u_k})_{1 \leq k \leq n} = (X_k)_{1 \leq k \leq n} \, .$$

Its covariance matrix is defined by the covariance function of the Gaussian field. Of course, this assumption is not always valid. It requires that the phenomenon under study is reasonably consistent with the local characteristics of the Gaussian field. But a wide range of textures can be synthesized depending only on an isotropic stationary

[1]This is an easy way to build a consistent spatial model, with a fixed univariate distribution and a fixed covariance function.

covariance[13]. Otherwise, other models should be used (e.g. Markov random fields, Boolean models).

The second assumption presupposes knowing of the sensing mechanism and/or available training data on which validation of the model can be performed. Examples can best illustrate this as follows:

- The k-th component of the vector of the observations may be the value of the variable under study at a particular point.

$$dF(z|x) = \delta_{\phi(x_1)}(z_1)\dots\delta_{\phi(x_p)}(z_p)$$

If the univariate transform ϕ is a one-to-one mapping, the observations fix the value of p components of the vector X: $x_k = \phi^{-1}(z_k)$. But if ϕ is not one-to-one, a set of possible values for X corresponds to each value of an observation. The model should define a probability on this set.

- The measurements are often noisy. A rather general model for noise is the following one[3, 6] (here the dimension of X and Z are equal). Given any particular value of X, the random variables $Z_1, \dots, Z_n$ are conditionally independent and each Z_i has the same conditional probability density function ($p.d.f.$) [2] $f(z_i|x_i)$, dependent only on x_i. Thus the conditional $p.d.f.$ of the observed records z given x is simply:

$$f(z|x) = \prod_{i=1}^{n} f(z_i|x_i)$$

This framework describes the usual additive and independent noise, $Z_i = \phi(X_i) + \epsilon_i$ where the $(\epsilon_i)_{1\leq i\leq n}$ are i.i.d. random variables whose common density function is h. In this case, $f(z_i|x_i) = h(z_i - \phi(x_i))$. But this noise model can describe more complex cases. For the micro-probe images, the photon-count Z_i is Poisson distributed with the actual concentration $\phi(x_i)$ as parameter.

- Indirect measurements of the phenomenon can be modeled by a sum of a discrete convolution and a noise : $Z_i = \sum_k \omega_i^k \phi(X_k) + \epsilon_i$. The random vectors X and $\epsilon = (\epsilon_i)_{1\leq i\leq p}$ are independent but correlations between components of ϵ are possible. Such models arise in remote sensing and seismic [16, 5]. If ϵ is a gaussian vector whose $p.d.f.$ is $g_\epsilon(\cdot)$.

$$f(z|x) = g_\epsilon\left((z_i - \sum_k \omega_i^k \phi(x_k))_{1\leq i\leq p}\right)$$

In the last two examples, the conditional $p.d.f.$ of the observations exists. Moreover, it often depends on a subset of the set on which the simulation has to be performed. We call A the set of the gaussian variables associated with the observations and S

[2] The positive measure, which defines the conditional $p.d.f.$, may be the Lebesgue measure on $\mathbb{R}$ or the indicator measure on $\mathbb{N}$.

the set of the other Gaussians to be simulated. Hence we have $A \cap S = \emptyset$, and we can rewrite:

$$X = (X_A, X_S) \quad \text{with} \quad \begin{array}{rcl} X_A & = & (X_i)_{i \in A} \\ X_S & = & (X_i)_{i \in S} \end{array}$$

The conditional *p.d.f.* of Z given x is therefore $f(z|x_A)$. Using Bayes formula, we can calculate the conditional *p.d.f.* of X given z :

$$g(x|z) = \frac{f(z|x_A)g(x_A, x_S)}{\int f(z|u_A)g(u_A, u_S)du_A du_S} = \frac{f(z|x_A)g(x_A)}{\int f(z|u_A)g(u_A)du_A} g(x_S|x_A) \qquad (1)$$

where $g(\cdot)$ denotes the *p.d.f.* of a standard Gaussian vector[3] and $g(\cdot|x_A)$ the conditional *p.d.f.* of a standard Gaussian vector given x_A. We conclude this section with few remarks on Formula 1:

- The conditional *p.d.f.* of the vector X given z is the product of two terms. Hence the conditional simulation is performed in two steps: first the conditional simulation of X_A is performed, then X_S is simulated with the simulated value x_A as a condition. As $card(S)$ is generally far larger than $card(A)$, and as the conditional *p.d.f.* of X_A given z is far harder to simulate than the conditional *p.d.f.* of X_S given x_A, we save a lot of computer time splitting the conditioning process in two steps.
- The conditional *p.d.f.* of X_S given x_A, $g(\cdot|x_A)$, is still a Gaussian *p.d.f.* Thus it can be simulated using a classical method as **conditional kriging** or **sequential Gaussian simulation** (section 3).
- The conditional *p.d.f.* of X_A given z is generally non Gaussian and we used the **Gibbs sampler** to simulate it (section 4).

3 Linear observations

In this section, we consider the case of linear observations: $Z = BX_A + R$ where B is a matrix and R a Gaussian vector[4] orthogonal to X. It is clear that (Z, X_S) is a centered Gaussian vector whose covariance matrix is

$$\begin{pmatrix} \Sigma_{RR} + B\Sigma_{AA}B' & B\Sigma_{AS} \\ (B\Sigma_{AS})' & \Sigma_{SS} \end{pmatrix} \text{ with } \begin{cases} \Sigma_{IJ} = E\{X_I X_J'\} \ , \ (I, J) \in \{A, S\}^2 \\ \Sigma_{SS} = E\{R R'\} \end{cases}$$

So the conditional distribution of X_S given z is a Gaussian distribution; its mean is the simple kriging of X_S given Z i.e.

$$X_S^{SK}(z) = z'(\Sigma_{RR} + B\Sigma_{AA}B')^{-1}B\Sigma_{AS}$$

[3]The covariance matrix and the dimension of the vector are defined by the letters used: $g(x_i)$ is the *p.d.f.* of the scalar Gaussian variable X_i and $g(x_A)$ is the *p.d.f.* of the Gaussian vector X_A.

[4]R may be zero, but the $B\Sigma_{AA}B'$ matrix should remain invertible.

and its covariance is

$$\Sigma_{SS} - \Sigma_{SA}B'(\Sigma_{RR} + B\Sigma_{AA}B')^{-1}B\Sigma_{AS}$$

Two methods are commonly used to simulate such a conditional distribution: sequential Gaussian simulation[1] and conditional kriging[12].

3.1 The sequential Gaussian simulation

In the Gaussian case, the sequential simulation is done as following:

- Select a component which has not been simulated yet, say X_i.
- Using all the components of (Z, X_S), measured or already simulated, compute the conditional distribution of X_i. It is a Gaussian distribution whose mean is the simple kriging X_i^{SK} and whose variance is $\sigma_{SK}^2(i)$. Hence the component X_i is simulated by
$$X_i = X_i^{SK} + \sigma_{SK}(i)U$$
where U is a standard Gaussian scalar.
- Continue until there are no more components to simulate.

Remark 1: This algorithm is equivalent to the method based on the Cholesky factorization of the covariance matrix of the vector (Z, X_S) [8] (the proof by induction is straightforward). But as algorithms to factorize an $n \times n$ covariance matrix are $O(n^3)$, the sequential Gaussian simulation may be extremely computationally heavy if n is large.

Remark 2: Using the kriging formalism, approximations are done to implement the sequential Gaussian simulation. At each step, instead of all the measured and the previously simulated components, only components belonging to a local neighborhood are used to compute the simple kriging. This approximation may lead to poor reproduction of the modeled covariance.

3.2 The conditional kriging

According to the conditional distribution of X_S given z, we can rewrite the conditional simulation as the sum of two terms:

$$X_S = X_S^{SK}(z) + U$$

where U is a residual orthogonal to the simple kriging $X_S^{SK}(z)$ — thus in the Gaussian case independent of it. As this residual is the estimation error, $X_S - X_S^{SK}$, it can be simulated as follows: the vector $(\tilde{Z}, \tilde{X}_S)$ is sampled and the estimation error computed $U = \tilde{X}_S - \tilde{X}_S{}^{SK}(\tilde{Z})$. But as the simple kriging is a linear estimator whose coefficients do not depend on the value of the data, the conditional simulation is the sum of the non conditional simulation and a linear combination of the covariances between an observation and the point to be simulated:

$$i \in S \quad X_i = X_i^{SK}(z) + (\tilde{X}_i - \tilde{X}_i{}^{SK}(\tilde{Z})) = \tilde{X}_i + \sum_{\alpha=1}^{p} \omega^\alpha Cov(Z_\alpha, X_i) \qquad (2)$$

as$X^{SK}(z) - \tilde{X}^{SK}(\tilde{Z}) = (z - \tilde{Z})'(\Sigma_{RR} + B\Sigma_{AA}B')^{-1}B\Sigma_{AS} = \Omega B\Sigma_{AS}$.

Remark 1: In this method[5], the conditioning process is split in two steps: a) the unconditional simulation of a Gaussian random function and b) the kriging process. For the first step, many methods are available: matricial, turning bands, ARMA and spectral methods (FFT or Shinozuka). Their respective efficiency depends heavily on the covariance model *and* the support S of the simulation. In the second step, the simple kriging is considered as an interpolator. But the interpolated variable is not the variable under study but rather the error between the conditional and the non conditional simulations.

Remark 2: The dimension of the vector Ω is that of the observation vector Z, generally far smaller than the dimension of the simulation X; it is computed only once. Moreover if the cross covariance between the observed variable and the variable to be simulated had a finite range, most terms in the sum 2 are equal to zero. In this case, a consistent moving neighborhood is defined.

4 Non linear observations

In this section, we assume that the conditional probability density function

$$\frac{f(z|x)g(x)}{\int f(z|u)g(u)du}$$

is non Gaussian. We present first the Gibbs sampler which was first proposed by S. Geman and D. Geman[11] in order to simulate Markov random fields but which can also be used to simulate the conditional *p.d.f.* of a Gaussian vector. Then this powerful tool is applied to three examples.

[5]This method is often referred to as the turning band method, but this designation may create confusion between a specific method to simulate a stationary random function and the conditioning process which relies on general properties of the simple kriging and the Gaussian model.

4.1 The Gibbs sampler

Let f be a *p.d.f.* on $\mathbb{R}^n$. We give up the idea of sampling f directly. So, starting with an initial simulation sampled from an arbitrarily chosen distribution, we modify it at random until we get a suitable vector. More formally, we construct a random process $(X^{(\nu)})_{\nu\geq 0}$ whose marginal distributions converge toward the joint distribution whose *p.d.f.* is f. Moreover we want this random process to be a homogeneous Markov chain, i.e.
(i) The conditional distribution of $X^{(\nu)}$ given its immediate predecessors $X^{(\nu-1)}$ is independent of all previous states.
(ii) It is defined by an initial distribution and a stationary transition probability.
The Markov chains are chosen because the computation of the series only requires that the current iteration need be stored in computer memory.

The transition probability of the Gibbs sampler is very simple. Each component of the vector is addressed in turn and when at the component i, the current value is replaced by a new one sampled randomly from the associated univariate conditional *p.d.f.* given all the other current components. When each x_i has been updated, one step of the Markov chain is complete.

4.1.1 The stationary distribution

To use the Gibbs sampler, we should first prove that the Markov chain is ergodic and f is its unique stationary *p.d.f.*, i.e. for any initial distribution, the distribution of the iterations $X^{(\nu)}$ converge toward the distribution whose density function is f. In all cases, we can prove that f is a stationary *p.d.f.* of the transition probability (proof in appendix). But when the configuration state is $\mathbb{R}^n$, no general results are available on the uniqueness of the stationary distribution of the Gibbs sampler. The studied cases are:

- f is the Gaussian *p.d.f.*[6] [15].
- *supp*$(f) = B$ is a compact set and $\mathcal{K}(x,y) > 0$, $\forall(x,y) \in B \times B$, where $\mathcal{K}$ is the density function of the transition probability[7][9].

Both are ergodic.

On the following two examples, we can see that non ergodic Gibbs samplers are not pathological cases. Figure 1 shows two 2D supports of the *p.d.f.* we would like to simulate using the Gibbs sampler. In the first case, the Gibbs sampler is ergodic, but in the second case it is not. As in the latter case, the chain remains in the

[6]But the Gibbs sampler is indeed a poor device to sample a Gaussian vector.

[7]This case is equivalent to the case of a Markov random field with a finite configuration state[11]

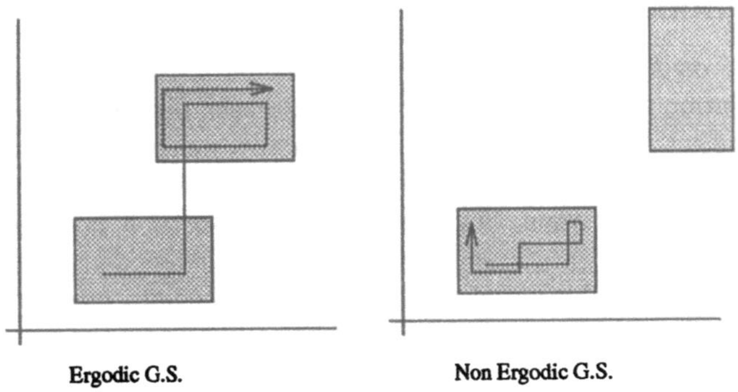

Figure 1: Two examples of 2-dimensional Gibbs samplers

connected component in which the initial vector has been simulated, the asymptotic distribution depends on the initial distribution. Allard[2] gives practical examples where such problems arise.

If $\mathcal{K}$ the density function of the transition probability is strictly positive on $I\!R^n \times I\!R^n$, an easy way to construct an ergodic chain is to substitute for the initial model, f, its restriction to a compact set B, $1\!\!1_B(x)f(x)/\int_B f(u)du$ [8]. If B is chosen so that $\int_B f(u)du \simeq 1$, the model is only slightly changed [9].

4.1.2 The speed of convergence

The theoretical criteria to forecast the speed of convergence are intractable for large vectors[10]. Thus we adopt an empirical criterion: on experimental studies[9, 10], statistics computed on the components of the iterations (e.g. mean and variance) seem to be good criteria for evaluating the time when the chain reaches its stationary state.

4.1.3 The Gaussian case

So far we have presented only general results on the Gibbs sampler. In the Gaussian case, the conditional $p.d.f.$ of X_i given the other components $\overline{x_i}$ and the observations z is proportional to

$$f(z|x_i, \overline{x_i})g(\frac{x_i - x_i^{SK}(\overline{x_i})}{\sigma_{SK}(i)})\frac{1}{\sigma_{SK}(i)} \; .$$

[8] $\mathbf{1}_B$ is the indicator function of the set B: $\mathbf{1}_B(x) = 1$ if $x \in B$ and 0 elsewhere.

[9] For an increasing sequence of compact sets $(B_n)_{n\geq 0}$ converging toward $I\!R^n$, we get an approximation of the model as, for any borel set C of $I\!R^n$, $\lim_{n\to+\infty} \int_{C\cap B_n} f(u)du \,/\, \int_{B_n} f(u)du = \int_C f(u)du$

[10] If $\delta = \min_{(x,y)\in B^2} \mathcal{K}(x,y)$, $|P\{X^{(\nu)} \in C\} - \int_C f(u)du| \leq (1-\delta|B|)^\nu$.

Therefore, we can write the algorithm as following:

$$X_i^{(\nu+1)} = X_i^{SK}(X_1^{(\nu+1)},\ldots,X_{i-1}^{(\nu+1)},X_{i+1}^{(\nu)},\ldots,X_n^{(\nu)}) + \sigma_{SK}(i)\,U_i$$

where X_i^{SK} is the simple kriging of X_i by the other components of the vector and $\sigma^2_{SK}(i)$ is the kriging variance. The residual U is sampled according a *p.d.f.* proportional to

$$f(z|x_i^{SK} + \sigma_{SK}(i)u, \overline{x_i})g(u)$$

It is a Gaussian density function modulated by a function depending on the observations.

4.2 Applications

We shall illustrate the use of the Gibbs sampler on an example. First we simulate a Gaussian image (200 × 200) whose covariance is spherical and isotropic. Its range is equal to 10 pixels (figure 2). From this, we get three observations:

a) a cut-off $(\mathbb{1}_{X_{i,j}\geq x_c})_{1\leq i,j\leq 200}$,
b) a Poisson noise $(\mathcal{P}_{\phi(X_{i,j})})_{1\leq i,j\leq 200}$,
c) a noisy 1D regularization $(\frac{1}{p}\sum_{k=0}^{p-1}\phi(X_{i+k,j}) + \sigma_\epsilon\epsilon_{i,j})_{1\leq i\leq 201-p,1\leq j\leq 200}$.

where ϕ is a lognormal anamorphosis ($\phi(x) = me^{\sigma x-\sigma^2/2}$ with $m = 5$ and $\sigma = .43$), $x_c = .67$ ($P\{X_{i,j} \geq x_c\} = .25$), ϵ is a centered Gaussian noise ($Cov(\epsilon_{i,j}, \epsilon_{i',j'}) = \delta_{j,j'}0.7^{|i-i'|}$ and $\sigma_\epsilon = 1.64$) and $p = 10$.

Conditional simulations based on these observations (figure 3) give good reconstructions of the initial scene. Now, we focus on some details of the computation. Later on, in order to simplify the notations, we shall refer to a pixel of an $n\times n$ image by a single index i running from 1 to n^2.

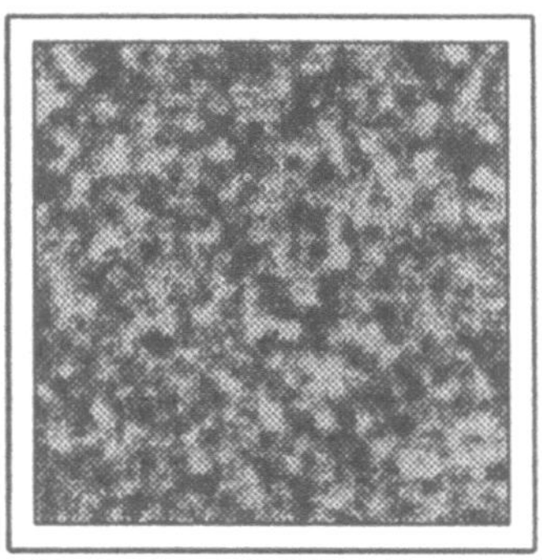

Figure 2: Initial simulation

a cut-off

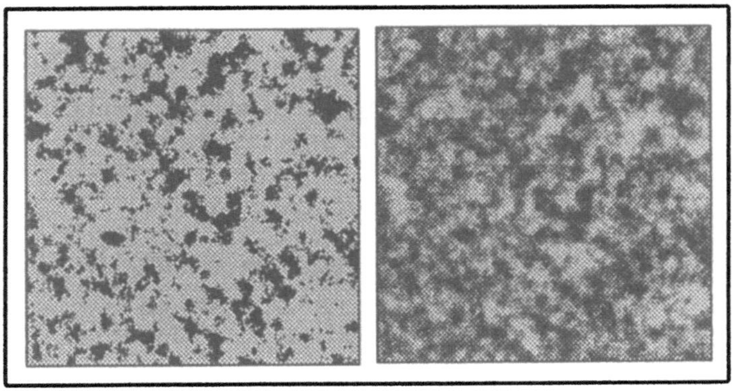

a Poisson noise

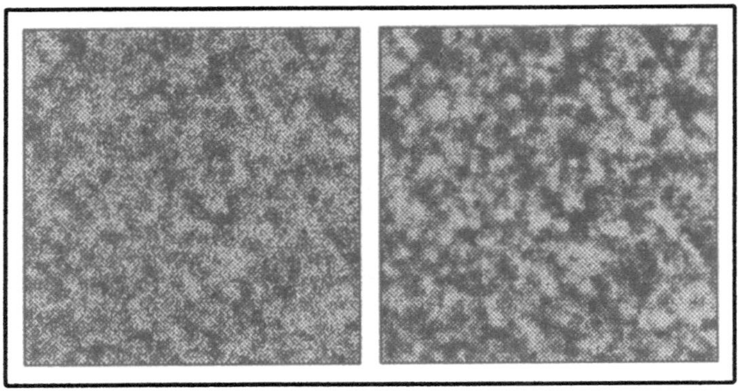

a noisy 1D regularization

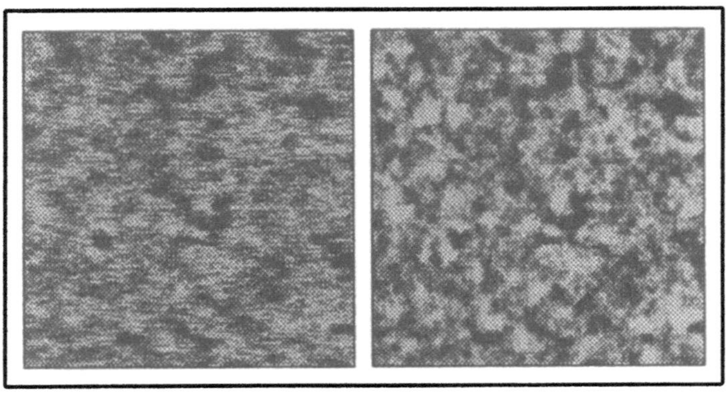

Figure 3: Three examples of conditional simulations

4.2.1 A cut-off

Let ψ denote $x \mapsto (\mathbb{1}_{x_i \geq x_c})_{1 \leq i \leq n^2}$. The observation is $z = \psi(x)$. As ψ is a step function, the conditional *p.d.f.* of X given z is

$$\frac{\mathbb{1}_B(x)g(x)}{\int_B g(u)du} \text{ with } B = \psi^{-1}(\{z\}) .$$

But B is a cartesian product of subsets of $\mathbb{R}$, $B = B_1 \times \cdots \times B_{n^2}$. So the conditional *p.d.f.* of the pixel x_i given the other pixels $\overline{x_i}$ and the observed indicator is

$$\frac{\mathbb{1}_{B_i}(x)g(\frac{x_i - x_i^{SK}}{\sigma_{SK}(i)})\frac{1}{\sigma_{SK}(i)}}{\int_{B_i} g(\frac{u_i - x_i^{SK}}{\sigma_{SK}(i)})\frac{du_i}{\sigma_{SK}(i)}}$$

This *p.d.f.* is simulated by an acceptance-rejection technique[10].

With this method, we can produce conditional simulations of random functions modeled as a transform by a step function[14, 9]. The Gibbs sampler can be used to simulate a Gaussian vector under the complex constraint "$X \in B$"[2]. Nevertheless, one should verify that $\int_B g(u)du > 0$ and that the Markov chain has an unique ergodic class.

4.2.2 A Poisson noise

For a Poisson noise, we assume the conditional independence of the values of the observed image given the true scene. Thus, the conditional *p.d.f.* of Z given x can be factorized: $f(z|x) = \prod_1^{n^2} f(z_i|\phi(x_i))$. The conditional *p.d.f.* of the pixel x_i, the values of the other pixels $\overline{x_i}$ and the observations z are known, is proportional to

$$f(z_i|\phi(x_i)) \; g(\frac{x_i - x_i^{SK}}{\sigma_{SK}(i)})\frac{1}{\sigma_{SK}(i)} .$$

This *p.d.f.* is simulated by an acceptance-rejection technique (see below).

This approach has been used to filter micro-probe images of chemical concentrations in steel samples. An approximation of the conditional expectation of non linear transforms of the concentrations can be computed by a Monte-Carlo method as $\lim_{n \to \infty} \frac{1}{n} \sum_1^n \psi(X^{(\nu)}) = E\{\psi(X)|z\}$ *a.s.* where the $(X^{(\nu)})_{\nu \geq 0}$ are the iterations of the Gibbs sampler[9, 7].

4.2.3 A Noisy 1D regularization

In a noisy regularization, the observations are $Z_\alpha = \sum_k \omega_\alpha^k \phi(X_k) + \sigma_\epsilon \epsilon_\alpha$. As X and ϵ are two independent Gaussian vectors, the conditional *p.d.f.* of X given z is proportional to

$$g_\epsilon((\frac{z_\alpha - \sum_k \omega_\alpha^k \phi(x_k)}{\sigma_\epsilon})_{1\leq\alpha\leq p})\ g_X(x)$$

where g_ϵ and g_X are the respective *p.d.f.* of the Gaussian vectors ϵ and X. We consider correlations between the components of ϵ, the likelihood function cannot be factorized and the computation of the conditional *p.d.f.* of X_i given $\overline{x_i}$ and z is harder.

Let A_i be the set of the observations which are modified by a change of the value of the component X_i: $A_i = \{\alpha, \omega_\alpha^i \neq 0\}$. This set may equal to A, but if the convolution function has a bounded support, A_i is strictly smaller than A. So the set of the observations can be divided in two subsets:
- Firstly, the components unmodified by the component X_i, A_i^c. But as z and $(x_j)_{j\neq i}$ are fixed, we can compute the value of the noise on these components,

$$\alpha \in A_i^c\ ,\ \epsilon_\alpha = \frac{1}{\sigma_\epsilon}\{z_\alpha - \sum_j \omega_\alpha^j \phi(x_j)\}\ .$$

- Secondly, the components modified by the component X_i, A_i. As we know the value of the noise for $\alpha \in A_i^c$, we can rewrite the noise of the components in A_i as the sum of two terms, the simple kriging by $(\epsilon_\alpha)_{\alpha\in A_i^c}$ and a residual: $\alpha \in A_i$, $\epsilon_\alpha = \epsilon_\alpha^{SK} - V_\alpha$. The value of the residuals is

$$\begin{aligned} V_\alpha &= \frac{1}{\sigma_\epsilon}\{\omega_\alpha^i \phi(x_i) - (z_\alpha - \sum_{j\neq i} \omega_\alpha^j \phi(x_j) - \sigma_\epsilon \epsilon_\alpha^{SK})\} \\ &= \frac{1}{\sigma_\epsilon}\{c_\alpha \phi(x_i) - d_\alpha\} \end{aligned}$$

Let $c_\alpha = \omega_\alpha^i$ ($\mathbf{c} = (c_\alpha)_{\alpha\in A_i}$) and $d_\alpha = z_\alpha - \sum_{j\neq i} \omega_\alpha^j \phi(x_j) - \sigma_\epsilon \epsilon_\alpha^{SK}$ ($\mathbf{d} = (d_\alpha)_{\alpha\in A_i}$). The vector of the residuals $V = (V_\alpha)_{\alpha\in A_i}$ is a centered Gaussian vector whose covariance matrix can be deduced from the covariance function of the noise. Q denotes the inverse of this covariance matrix. The *p.d.f.* of these residuals is proportional to

$$\exp\{-\tfrac{1}{2}V'QV\} \propto \exp\{-\tfrac{1}{2}(\frac{\phi(x_i) - a}{b})\}$$

with

$$\begin{cases} a = \frac{\mathbf{c}'Q\mathbf{d}}{\mathbf{c}'Q\mathbf{c}} \\ b = \frac{\sigma_\epsilon}{(\mathbf{c}'Q\mathbf{c})^{\frac{1}{2}}} \end{cases}$$

Finally, the conditional *p.d.f.* of X_i given z and $\overline{x_i}$ is proportional to

$$\exp\{-\tfrac{1}{2}(\frac{\phi(x_i) - a}{b})\}\ g(\frac{x_i - x_i^{SK}}{\sigma_{SK}(i)})\frac{1}{\sigma_{SK}(i)}$$

Here too, this *p.d.f.* is simulated by an acceptance-rejection technique.

We can give a simple example of the coefficients a and b: if the observations are $Z_\alpha = \phi(X_\alpha) + \sigma_\epsilon \, \epsilon_\alpha$, where ϵ is a noise without correlation, which is the simplest example of the general case presented above, $a = z_\alpha$ and $b = \sigma_\epsilon$.

The noisy 1D regularization model is used for post stack seismic trace[16]. The observed amplitude is the result of a convolution of the reflexion coefficients in time by the signal plus a noise. Thus this algorithm could be used to constrain stochastic images to seismic data. Bortoli et al.[5] propounded a similar algorithm but using the Gibbs sampler we can introduce correlation in the noise and ensure the convergence.

4.2.4 Some acceptance-rejection techniques

In the last two examples, the conditional *p.d.f.* that we have to simulate is proportional to the product of two terms :

$$h \circ \phi(x) \; g(\frac{x - x_i^{SK}}{\sigma_{SK}(i)}) \frac{1}{\sigma_{SK}(i)} \; .$$

One technique for simulating such a density function when ϕ is a step function[11], $\sum_k \phi_k \mathbb{1}_{b_k}(x)$ where the $(B_k)_k$ are disjoint intervals on the real line, is given in [9]. We can rewrite $X = x_i^{SK} + \sigma_{SK}(i)U$. The *p.d.f.* of the residual U is a mixture of truncated Gaussian variables:

$$\sum_k \varpi_k \frac{\mathbb{1}_{B'_k}(u) g(u)}{\int_{B'_k} g(v)dv} \text{ with } \left\{ \begin{array}{lcl} B'_k & = & \frac{1}{\sigma_{SK}(i)} (B_k)_{-x_i^{SK}} \\ \varpi_k & = & \frac{h(\phi_k) \int_{B'_k} g(u)du}{\sum_{k'} h(\phi_{k'}) \int_{B'_{k'}} g(v)dv} \end{array} \right.$$

Another technique is to compute $C = \max_x\{h \circ \phi\}$ and apply the standard acceptance-rejection technique:

(i) Simulate (X, U) according the *p.d.f.* $g(\frac{x - x_i^{SK}}{\sigma_{SK}(i)}) \frac{1}{\sigma_{SK}(i)}$ and $\mathbb{1}_{]0,1[}(u)$.

(ii) If $CU \leq h \circ \phi(X)$ the value of X is kept, otherwise return to (i).

If ϕ is a positive and strictly increasing mapping, the computation of $\max h \circ \phi$ is straightforward:

- for the Poisson noise,

$$\max\{e^{-\phi(x)}(\phi(x))^z\} = \left\{ \begin{array}{ll} 1 & \text{if } z = 0 \\ e^{-z}(z)^z & \text{if } z \geq 1 \end{array} \right.$$

- for the noisy regularization, $\max\{e^{-\frac{1}{2}(\frac{\phi(x) - a}{b})}\} = 1$.

[11] or can be approximated by a step function

Appendix

f is a stationary probability density function of the Gibbs sampler.

Proof: Let $\mathcal{K}$ be the density function of the transition probability. It is clear that,

$$\mathcal{K}(x,y) = \prod_{i=1}^{n} f(y_i|y_1,\ldots,y_{i-1},x_{i+1},\ldots,x_n) \; .$$

When $\forall i \in \{1,\ldots,n\}$, the conditional *p.d.f.* is

$$f(y_i|y_1,\ldots,y_{i-1},x_{i+1},\ldots,x_n) = \frac{f(\ldots,y_i,x_{i+1},\ldots)}{\int f(\ldots,y_{i-1},u_i,x_{i+1},\ldots)du_i} \; .$$

The *p.d.f.* f is stationary if $\int f(x)\mathcal{K}(x,y)dx = f(y)$. By Fubini's theorem,

$$\begin{aligned}
\int f(x)\mathcal{K}(x,y)dx &= \int dx_n f(y_n|y_1,\ldots,y_{n-1}) \int dx_{n-1}\cdots\int dx_1 f(y_1|x_2,\ldots,x_n)f(x) \\
&= f(y)\int \frac{dx_n f(y_1,\ldots,y_{n-1},x_n)}{\int du_n f(y_1,\ldots,y_{n-1},u_n)}\int dx_{n-1}\cdots \\
&\quad \int \frac{dx_2 f(y_1,x_2,\ldots,x_n)}{\int du_2 f(y_1,u_2\ldots,x_n)}\int\frac{dx_1 f(x_1,\ldots,x_n)}{\int du_1 f(u_1,x_2,\ldots,x_n)} \\
&= f(y)
\end{aligned}$$

■

References

[1] F. Alabert. *Stochastic imaging of spatial distributions using hard and soft data.* Msc. thesis, Stanford U, 1987.

[2] D. Allard. Simulation of a geological lithofacies with respect of connectivity information using the truncated gaussian model. In M. Armstrong, editor, *Workshop on geostatistical simulations*, volume 1, pages 1–2, Dordrecht Holland, 1994. Kluwer Academic Press.

[3] J. Besag. On statistical analysis of dirty pictures. *J. R. Statist. Soc. B*, 48:259–302, 1986.

[4] J. Besag, J. York, and A. Mollié. Bayesian image restoration, with two applications in spatial statistics. *Ann. Inst. Statis. Math.*, 43(1):1–59, 1991.

[5] L.J. Bortoli, F. Alabert, A. Haas, and A. Journel. Constraining stochastic images to seismic data. In A. Soares, editor, *Geostatistics TRÓIA'92*, volume 1, pages 325–337, Dordrecht Holland, 1993. Kluwer Academic Press.

[6] C. Daly. *Applications de la Géostatistique à quelques problèmes de filtrage.* thèse de docteur en géostatistique, ENSMP, Fontainebleau, France, 1991.

[7] C. Daly. Filtering non-linear functions. In A. Soares, editor, *Geostatistics TRÓIA'92*, volume 1, pages 61–72, Dordrecht Holland, 1993. Kluwer Academic Press.

[8] M. W. Davis. Production of conditional simulation via lu decomposition of the covariance matrix. *Math. Geol.*, 19:91–98, 1987.

[9] X. Freulon. *Conditionnement du modèle gaussien par des inégalités ou des randomisées.* thèse de docteur en géostatistique, ENSMP, Fontainebleau, France, 1992.

[10] X. Freulon and Ch. de Fouquet. Conditioning a gaussian model with inequalities. In A. Soares, editor, *Geostatistics TRÓIA'92*, volume 1, pages 201–212, Dordrecht Holland, 1993. Kluwer Academic Press.

[11] S. Geman and D. Geman. Stochastic relaxation, gibbs distribution and the bayesian restoration of images. *I.E.E.E. transactions : Pattern Analysis and Machine Intelligence*, 6:721–741, 1984.

[12] A. G. Journel. *Simulations conditionnelles de gisements miniers.* thèse de docteur-ingénieur, Université de Nancy-I, Nancy, France, 1977.

[13] Ch. Lantuéjoul. Non conditional simulation of a gaussian random function by the turning band method. In M. Armstrong, editor, *Workshop on geostatistical simulations*, volume 1, pages 1–2, Dordrecht Holland, 1994. Kluwer Academic Press.

[14] G. Matheron et al. Conditional simulation of the geometry of fluviodeltaic reservoirs. *SPE*, 1987. 62nd Annual Conference, Dallas, Texas.

[15] L. Younes. *Probleme d'estimation parametrique pour les champs de Gibbs Markoviens – Application au traitement d'images.* Thèse, Université d'Orsay, 1988.

[16] Z. Zhang and A. Galli. Getting better quality seismic sections. In A. Soares, editor, *Geostatistics TRÓIA'92*, volume 1, pages 285–297, Dordrecht Holland, 1993. Kluwer Academic Press.

DISCUSSION FOLLOWING SESSION NO 2

Chairman: Alain Galli

Papers: Hegstad; Freulon

Gomez: Could you please explain for a layman where you would get all these π^*? How would you put this into practice to say simulate a sand shale?

Omre: The details are given in the paper.

Lantuéjoul: I have a problem with the simulated annealing method in the sense that it has been devised for optimising functions, whereas you are using it to simulate random functions. So could you tell me where the random part appears in the algorithm that you have presented?

Hegstad: It is in the construction of the Markov chain, in the transition probabilities.

Lantuéjoul: So randomness is used to find the minimum of the function. You also give the initial states; you start with a random function. So randomness is involved in two places: firstly, in the initial state and then to run the optimization algorithm. Earlier this morning Henning Omre said that it was very important to specify the underlying model. So I would like to ask you what it is here.

Hegstad: The model of the reservoir is given by the parameters η, etc.

At that point, the chairman stopped the discussion in order to allow enough time for the next presentation. After that was finished, the discussion started up on Freulon's paper.

Jeulin: The technique described in your presentation looks like an alternative to filtering. Concerning the filtering, I have two questions. Firstly, in a real situation, you start out with a noisy image, and you have to go back to the underlying properties of the image which are not here. For example the histogram of the noisy image is not the same as that of the real one. Could you comment on this? Secondly, could you comment on the signal to noise ratio that you get, compared to other filters such as cokriging.

Freulon: In answer to your first question, there has been a lot of work on the inference of the underlying parameters. For example, for the Poisson

M. Armstrong and P. A. Dowd (eds.), Geostatistical Simulations, 72–77.

noise case, I used the results given in Colin Daly's thesis. In more general cases, there are some problems but I have not looked at them in detail. It was just to see how the simulation algorithm worked.

Jeulin: But you tried to replace the initial gaussian random function by something else.

Freulon: First I will reply to your second question on the filtering property of the conditional simulation before coming back to this point. I do not think that this is a filter. To construct a filter you would have to take the average of lots of simulations to get the conditional expectation. Or maybe we could change the temperature via simulated annealing to get an a posteriori maximum. We would then get a filter. But it was not the purpose of the method which was to produce images that respect certain constraints and with given statistical properties.

Jeulin: But it is very similar to a filter because...

Freulon: And of course, in this case, they are not optimal. To get an optimum filter, you have to add something. First I started using the Poisson noise case to construct the conditional expectation and Colin Daly suggested it could be a good by−product of this method.

Jeulin: It is not gaussian...

Freulon: Now coming back to your other point, instead of using a gaussian variable, I started with a mosaic model which is quite different and I compared it with a linear filtering given by disjunctive kriging. It worked as well as the other methods. So it seems quite robust. But you have to infer the right covariance. Of course the conditional simulation will be multigaussian rather than mosaic.

Daly: One of the really interesting things in this paper is the deconvolution because that has got huge potential applications for inverting seismic data. In practice seismic data that you see, has lost low frequencies. The convolution takes place not on the acoustic impedance but on the reflectivity, essentially on the derivative. Would your technique work just as well for inverting in that case?

Freulon: I think that you can compute the conditional distribution of the observed values given the gaussian one. It is more complicated but with a little bit more work it should be possible.

The chairman then closed the discussion just on Freulon's paper, in order to widen the scope of the discussion. To start it out, he asked Lantuéjoul to repeat the question that he had asked earlier.

Lantuéjoul: My questions were: "Where is the randomness?" and secondly "Where do you introduce randomness into the optimization algorithm?" Let me put it differently. The result you get from simulated annealing should probably depend on the initial state.

Hegstad: It is a Markov chain. The limiting distribution is independent of the initial one.

Omre: It converges in probability, independent of the initial state. You asked where the randomness is. We have a high dimensional model and our constraints are of much lower dimension. There are many optimums. The important result in Geman and Geman is that it is uniform, it reaches solutions uniformly over the optimums and that creates randomness and that is exactly what you go on in your Gibb's sampler too.

Lantuéjoul: Suppose that you want to carry out a simulation with 4 facies. Suppose that in the initial stages, you consider a simulation with 6 or 7 facies. What will happen to the 2 or 3 additional facies that should not exist?

Omre: You have to start out from an initial case where you have a positive probability. You cannot start out with an impossible initial case. That is one constraint.

Lantuéjoul: Yes but what is a positive probability? How do you measure it?

Omre: Let us sit down and read through the paper. Your questions are very speculative. The important aspect is that it is multi–optimum and from Geman and Geman's results, you reach it uniformly. And it is the same thing that you draw on in your Gibb's sampler.

Freulon: The result on the asymptotic distribution given by Geman and Geman is uniform on the possible maximum states. And it is a conditional model of what you want to simulate. I think it is defined only by the constraints and not by a probability model.

Galli: Is it the same question?

Daly: Maybe it is. By changing the probability structure temporarily whilst annealing in the constraints, it is no longer obvious that in the case of several connected components for the constraints, the annealing will converge to one of these with the appropriate initial probabilities. It seems that we would need some theorem on the uniform convergence of the algorithm. I am not describing this very well. Do you see what I am saying?

Omre: I understand your question. We think that what we are doing is correct but this is not in the paper.

Lajaunie: One point to note is that this is obviously one way to specify the target function, the Φ function. If it happens that you have a unique absolute minimum, then you have no randomness at all. So have you investigated the set of absolute minima for the function that you are specifying. You said that there were multiple minima, but are you sure about this?

Omre: We have not studied this in particular but in the cases we have worked on, very often you have a lot of degrees of freedom so it is easy to satisfy the constraints. In most of the cases I have met in practice you have many solutions that can live up to the constraints. We have not studied what happens in the limit when you have many constraints. Did I answer your question?

Lajaunie: In my opinion there is not enough richness in the set of minima to provide enough for a consistent simulation technique.

Omre: Think about a gaussian random function being tied up in one single observation. Then we are back to our initial situation.

Lajaunie: We have agreed that this Markovian simulation technique is just an intermediate used for calculation, to reach the minimum of Φ. So it is not a source of randomness in the result.

Omre: To be honest, I believe that Chris Farmer and Clayton Deutch who presented papers on this have not looked carefully enough into the basic theory; so they explain it in a fairly obscure way. But we are trying to look into it in more detail to see why it works. Clearly techniques like steepest decline cannot be used here. So I would suggest that there is potential here, and also a lot of questions to be solved. Here we have seen Xavier Freulon using a Gibb's sampler. He avoids the problems of determining complex conditional distributions by using a form of Hastings' algorithm. So there is a lot of richness in the model. We are not able to answer most of the questions you can pose but I insist that there is potential in this approach. I insist that Deutch and Farmer did something smart and that this method can be used in an efficient manner in complex models.

Jeulin: One of the points of this model is to respect constraints. Presumably the set the maxima depends on the choice of the constraints, and of course on the cost function. Did you try different cost functions for the same constraints? Did you compare the simulations? Could you give us some guidelines for choosing the cost functions?

Omre: Up till now the space of models that we work in has been so limited. So we have been trying to expand this space of models that we can simulate properly. And of course there are many convergence problems, severe convergence problems. But we have not yet explored the model sufficiently to be able to answer your questions. But we would like to.

Jaquet: I am not a theoretician. So here is a more practical question. Could you comment on the underlying concepts of your model and the truncated gaussian? Suppose I want to do a simulation. How should I choose between the two techniques. What should you look at? What assumptions should be checked?

Omre: There are two main classes; one is sequential simulation algorithms that can be used for gaussian functions, and then you have the iterative ones. If you imagine a grid system, you can simulate one point, then another point conditional to that and so on. You can do that because you can compute all the conditional distributions analytically. If you have a more complicated model than a gaussian, you may be able to specify some inter–relations but you cannot calculate the conditional distribution given two shapes where to put the third one. So we have to put them in, in a random manner and then reshuffle them iteratively and then it will converge to the correct probabilities. So it is a way of being able to simulate from an arbitrary n dimensional distribution. Can you see what I mean?

Jaquet: Sort of.

Haas: Concerning all the methods with cost functions, the interesting point is that you can mix lots of different types of conditions, but if your constraints are not mutually consistent you can get some very surprising results. For example, looking at Clayton Deutch's work, you can see the phenomenon of compensation of permeabilities around the well because the different constraints were the distribution of permeabilities and a given permeability for the well. And as it happens, the permeability at the well was not in the distribution. So by optimizing, you obtain something that is completely unrealistic.

Omre: First I think that we should avoid the word "optimization" and talk about "conditioning by annealing". Secondly, I agree with you. If you have such a flexibility in putting in conditions they may conflict completely, if you require them to be exact. But we work on more flexible ones. The method allows room for constraints with any probability including certainly so these conflicting situations are fewer.

Haas: The main problem is to choose a good set of constraints. Suppose you want to respect the variogram exactly. But no one knows the true variogram.

Omre: I agree. It is stupid to try to reproduce the variogram which is a model parameter. You should try to reproduce something that you can actually observe like a value of a variable. So I agree completely on that but the model is very rich.

As there were no more questions, the chairman closed the session.

CONDITIONAL SIMULATION FOR MINING: PRACTICAL IMPLEMENTATION IN AN INDUSTRIAL ENVIRONMENT

Peter J Ravenscroft
RTZ Technical Services Ltd
Castlemead, Lower Castle Street
P.O. Box 50
Bristol BS99 7YR
England

This paper provides an insight into some of the practical aspects of introduction of the conditional simulation approach into a major international mining company. Although geostatistical theory and academic understanding of advanced techniques such as conditional simulation may be well developed, their practical implementation in the mining industry provides new challenges. While it is in no way suggested that the speed of academic development be changed, it is hoped that some of the experience related in this paper may provide new perspectives to those involved in theoretical research and development, and ensure that practical solutions are available for the practical problems faced by industry.

An initiative was made in 1990 by RTZ Technical Services to examine the technique of conditional simulation as an additional tool to enhance the wide range of reserve estimation and mine planning technology in use throughout the RTZ Group. A study was to be conducted of the different simulation techniques currently available, with a view to selecting the most suitable method of conditional simulation for practical implementation. A software system was to be developed to enable this technology to be readily available and compatible with other estimation and mine planning software, and a number of case studies on current mining operations were to be undertaken in order to validate the technique and demonstrate its application.

This implementation study was completed over a period of two years, and this paper discusses some of the more important difficulties faced and problems resolved.

INTRODUCTION

Conditional simulation has been an accepted technology in academic circles and in some industrial sectors for nearly twenty years. However, although initially developed to address mining problems, its use in the mining industry has seemingly been limited to a few case studies and, to the author's knowledge, very few live applications. After an initial period of interest when some case studies were presented, very little was published on the application of conditional simulation in the mining industry.

There are many possible reasons for this, but the most likely are the early high cost of

M. Armstrong and P. A. Dowd (eds.), Geostatistical Simulations, 79–87.

conditional simulation in terms of computer requirements, and the perceived complexity of the method in a traditionally conservative industry.

However, with increasingly cheap computer power and the development of more simple simulation algorithms, there is renewed interest in the routine application of the method, as can be seen by some of the more recent publications on practical mining applications of conditional simulation (Srivastava et al., 1992; Nowak et al., 1992).

It was in this changed context that an initiative was made by RTZ Technical Services Ltd in 1990 to undertake a comprehensive study into the state of the art of conditional simulation, with a view to demonstrating the potential of its application and introducing it as an enhanced tool for reserve evaluation and operational grade control within the RTZ mining group.

In addition to the expected potential, this development programme has shown up a number of difficulties and limitations in the implementation and validation of the technique. The discussion of some of these in this paper is intended to stimulate interest from those involved in further research and development and to provide new perspectives on implementation of complex technology such as conditional simulation in a practical environment.

ACCEPTANCE AND CREDIBILITY

The introduction of earlier geostatistical technology, such as kriging, into the mining industry in the 1970's and 1980's faced the usual problems of scepticism and resistance to change which are normal with any new methods. Acceptance of linear geostatistics is still not automatic in some parts of the industry, and the routine use of geostatistical methods is often less widespread than might be thought.

The credibility of geostatistics in some mining companies has even been lessened by the experience gained in historical applications. New techniques provided by geostatistics were embraced fervently in the belief that some of the benefits would be immediate and as lucrative as claimed. Unfortunately some earlier proponents of the technology over-sold many of its virtues, and a number of mining companies have sad experiences of burning their fingers on false hopes and unachievable promises.

Introduction of more advanced methods such as conditional simulation in the 1990's must be seen against this background of continued, and in some cases heightened, scepticism. New and unfamiliar technology can only reach an acceptable level of credibility once proven in practice, and in the implementation programme in RTZ emphasis was placed both on validation of the technique in a number of varying geological and operational situations and on demonstration of some of the concrete advantages that could be attained.

CHOICE OF SIMULATION METHOD

When the conditional simulation approach was first developed in the 1970's (Journel, 1974) the method proposed was that of turning bands, with the familiar creation of a non-conditional simulation model and subsequent conditioning to the known data values. While this approach is still very much accepted technology and is still in wide use, a number of alternative methods of conditional simulation have been developed over the years.

Of these, the most promising from the point of view of practical implementation have been matrix decomposition methods (Alabert, 1987), autoregressive methods (Boulanger, 1988) and sequential methods (Gomez-Hernandez and Srivastava, 1990; Isaaks, 1990). All have suffered from distinct limitations, especially the restrictions in size of models which can be generated by the matrix decomposition approach or the difficulty in conditioning inherent in other methods.

Other algorithms, which for the moment can possibly only be considered as academic developments yet to be proven in practical implementation, include the annealing approach (Farmer, 1989) and substitution models (Lantuéjoul, 1992).

In selecting a method for practical implementation and ongoing application in a mining environment the main decision criterion was the proven applicability of the method and the degree to which successful implementation in industrial applications had already been achieved. Importantly, the method also had to be *conditional*: although non-conditional techniques may be of interest to academics, they have almost no relevance to real-world applications and real-world problems.

In addition to this, several other criteria were defined:

1. *Simplicity of method:*

   - simplicity of mathematical algorithm and ease of understanding of concepts for non-mathematically trained geologists and engineers;
   - relation to known geostatistical techniques, such as kriging, to enable a logical sequence in comprehension of method;
   - ease of software implementation to allow for portability of software and ongoing development and support by non-specialists.

2. *Robustness of technique:*

   - sensitivity to changes in parameters and degree of repeatability;
   - capacity for generating consistent results in an uncontrolled application (i.e. suitability for "black box" implementation in an operational environment).

3. *Speed of algorithm.*

4. *Size of simulation models:*

   - ability to create very large (tens of millions of nodes) models for simulations of entire deposits;
   - ability to create small (hundreds of nodes) individual models for simulations of blasts or stopes;
   - ability to generate multiple simulations in parallel.

On the basis of the main criterion of proven applicability, initial simulation work was performed in RTZ using the classical turning bands approach, and a number of pilot studies were done using an in-house version of the technique. However, several limitations became apparent, mostly related to the other criteria outlined above.

The first of these was the complexity of the approach and the difficulty in explaining the underlying algorithms to practical geologists and engineers. As mentioned earlier, the times of being able to introduce advanced geostatistical technology easily into a mining company are past: most modern mining management has an understanding of geostatistical terminology such as variograms and kriging, and is aware of the pitfalls of accepting blindly any new variations on more traditional approaches.

However the mathematical complexity of creating one-dimensional simulations on lines in a strange spatial configuration, projecting these into three dimensions and then kriging everything to ensure a good fit to the data was clearly too convoluted to be accepted easily. Similarly, this complexity required a full understanding of the intricacies of the method for anyone involved in software development and maintenance, which is again not realistic in a large and diverse mining company.

Secondly, in trial runs of the turning bands method, the sensitivity to input parameters became apparent. The known limitations of number of lines in space were obviated by using large numbers of lines (up to 360 lines in two-dimensional applications and up to 1000 lines randomly distributed in three-dimensional space), but sensitivity to the degree of discretisation on lines was also noted. It is likely that further test work in this area would have established an optimal approach, but the difficulties in implementation and understanding were thought to outweigh the possible benefits to be obtained.

The other simulation approach which appeared to meet the criteria discussed above was the sequential method developed more recently and well explained by several authors (Isaaks, 1990; Verly, 1992)[1]. In particular, the simplicity of the approach and its close relationship to the accepted technology of kriging allowed for easier acceptance and much simpler software implementation. Extensive trials were made of the performance of the method, and satisfactory results were readily obtained with a minimum of test work on sensitivity to parameters and differences in simulation environments.

The use of the sequential gaussian approach was favoured for most mineral grade modelling applications, with the indicator approach being used for more complex distributions or specific problems. In addition, with a simple application of indicator simulation to model spatial attributes such as geological rock types, the sequential simulation family offered a quick and robust solution to most simulation requirements for mining applications.

The decision was thus made to use sequential simulation as the basis for all conditional simulation work in the development exercise, and the technology was accepted as the most practical way to provide a simple, robust, and manageable implementation of conditional simulation for practical application.

---

[1] In fact, the method was used as early as 1974 in practical application, but was only recognised as being theoretically sound more recently (H.M. Parker, personal communication)

SOFTWARE AND HARDWARE IMPLEMENTATION

An important requirement for development of a software system to provide conditional simulation capabilities was portability of software and ease of implementation on a variety of platforms. Although more efficient performance could probably be obtained from the 'C' programming language, it was decided to stay in the more familiar FORTRAN environment for which a greater level of support exists in the industry.

The basis for the simulation algorithms used was from the GSLIB software library (Deutsch and Journel, 1992), but a customised version involving optimised file storage facilities and substantial support software was developed.

At the outset, initial development work was done using DOS-based PC's, with the intention to transfer the software to more powerful UNIX workstations for actual application. However, the rapid expansion in PC power, even during the time of development of the system, soon showed that adequate capacity was available on a PC platform and that the advantages of staying in the DOS environment outweighed the complexities of porting to less universal workstation machines.

At the time of writing, PC machines with a 66 MHz processor and memory capacity of 64 Mb provide ample power for conditional simulation work. A recent simulation study on a gold deposit involved the simulation of around 27 million nodes with six different attributes, and the entire computation was completed in less than 24 hours.

VALIDATION OF CONDITIONAL SIMULATION

As mentioned earlier, one of the critical aspects of introducing conditional simulation technology into a practical environment was the validation of the method and the demonstration of its application in a number of case studies. Although validation of such a technique should be relatively simple in an industry where vast amounts of sampling data are recorded, it was actually found that definitive proof of the method is an impossible goal, and the identification of what one is actually trying to validate is extremely complex.

Simplistically, initial validation exercises followed the lines of using wide-spaced exploration drilling data from a given deposit to create a detailed simulation model, and then checking the characteristics of this model against subsequent close-spaced blasthole sampling from the same deposit. With the wide variety of orebodies mined within the RTZ group and the access to almost limitless historical sampling information, it was considered a straightforward task to produce a variety of case studies fully validating and demonstrating the method.

However problems immediately became apparent in determining which information was the most representative of the actual mineralisation within the deposit: was exploration drilling data, with careful sampling and assaying techniques but much sparser coverage, more representative than blasthole sampling data, with less precision in sampling but with much larger samples and denser coverage?

Although techniques are available, and were used, to take into account different size of sample and even the sampling precision of blastholes, the application of such adjustments to simulated values followed by validation with actual blastholes becomes a self-fulfilling prophecy. Similarly, the simulation of blasthole-type values is in itself a somewhat pointless exercise: in most applications of conditional simulation the objective is the creation of a model of what is really in the ground and not of what we see from imperfect sampling techniques.

This leads to another philosophical question: when creating initial simulation models based on some form of sampling data, should the variogram used for the simulation include the portion of the nugget effect which is attributable to sampling error, or should the simulation be made to characterise "in-situ nugget effect" only?

Conversely, it might be considered that a better measure of what was in the ground might be obtained from using production data of grades and tonnages delivered to the mill. Again, apart from problems of dilution and inaccuracies in sampling and measurement, the scale at which such data is available precludes any successful determination of the short-scale level of variability which conditional simulation was designed to model. Most mines perform some form of blending, and even in the rare case of being able to isolate plant feed from one particular face or stope, the exact spatial location of the source of material would not be precise enough for the purpose of the validation required.

When all aspects of the objectives of the simulation approach and the limitations in being able to test the results are considered, it becomes apparent that the validation of such a model is inherently impossible. In mining applications, the model should be constructed in such a way as to represent as well as possible the in-situ spatial characteristics of the attribute being modelled. No measure of these in-situ characteristics in a real deposit can ever be obtained, since we are only capable of imperfect sampling and measuring. We are thus left with the need for intrinsic belief in the model, and the kind of "act of faith" discussed much more eloquently in the past (Matheron, 1978).

As emphasised earlier, this need for mining management to believe implicitly and accept a technique which can not be validated in any normal engineering manner is probably one of the greatest weaknesses and limitations of the approach.

APPLICATIONS

In order once again to put into perspective certain practical requirements of conditional simulation in industrial use, it is useful to discuss some of the applications intended for the method. These can be divided into two main areas based on the two aspects of characterisation of variability provided by conditional simulation:

- modelling of local short-scale variability in a way not provided by any interpolation technique;

- generation of equi-probable alternative images to allow for sensitivity and risk analysis.

The first of these, providing models of mineral grades, geological features or any other attributes on a detailed local scale, allows for spatial modelling of realistic variability rather than the artificially smoothed representations produced by any other technique. Block models with any size of block can be produced, and local frequency distributions can be obtained directly from the model.

The first application of simulation in RTZ was thus a simple approach to problems of change of support and recoverable reserve estimation, using a simulated model to provide transformations of grade/tonnage distributions from large kriged blocks to smaller selective mining units, and to simulate blasthole kriging and misclassification during grade control. The choice of conditional simulation to provide these results, as against more specialised forms of non-linear kriging, was made because simulation produces a model which is more compatible with traditional mine planning technology. Previous experience has shown that dealing with distributions rather than numeric values is difficult to implement, and classical mine planning philosophy can not assimilate such information easily. In addition, it has been found that conditional simulation provides a means of incorporating geological control and local data conditioning in a more simple and robust way than other alternative techniques of recoverable reserve estimation.

Another application of the modelling of local variability is in providing predictive models of variability in mill feed from an operating mine. These can be models of the characteristics of primary minerals, such as the grade or quality of material produced from an operation where tight restrictions are placed on product specifications and where stockpiling and blending are required to ensure a constant final product quality. Alternatively they can be models of the variation in quantity of deleterious elements which adversely affect the performance of the treatment process, and which again must be blended with other material to minimise their effect. In both cases the simulation model might be used in conjunction with a time/process simulation to determine blending requirements or to optimise stockpile size.

The applications discussed above normally require only one or two realisations of the simulation, and would typically involve simulating the entire deposit or at least a large representative portion of the deposit. They, as in any other mining application, obviously need to be conditional and to respect all known information about the orebody or about the mining operation.

The second area of application, as outlined earlier, is the use of multiple simulations to provide a measure of variability *between* simulations as against the previous focus on variability *within* simulations. This approach characterises the uncertainty of the modelling process, and has clear application in sensitivity and risk analyses. It is important to emphasise, however, that the variability identified by the simulation is limited to the intrinsic variability of the process being modelled, and can in no way encompass such external factors as sampling error and bias, core recoveries, or even inconsistencies in the underlying geological model.

Applications currently being considered in this area include a risk analysis approach to mine scheduling, where the robustness and flexibility of a given mine schedule can be tested by applying it to a number of alternative simulation models.

Similarly, the variability can be used as additional information when deriving the initial mine schedule: mining engineers are showing an increasing willingness to incorporate probabilistic information into decision making, and the selection of a block to schedule for mining can be based on the probability of it being above cut-off rather than on its absolute value. This application is being considered for implementation both in a project planning environment, and in an operational situation where probabilistic information, combined with an expert systems approach, can provide more effective operational scheduling.

Again, the results produced using the simulation approach could possibly be achieved by alternative methods of predicting variability and confidence limits by more mathematical techniques, but conditional simulation is once more favoured as producing results in a readily comprehensible format and allowing for direct visual characterisation of variability. Although it could be argued that the approach is highly computer-intensive and slow, these factors are becoming of less concern as cheap computing power becomes increasingly easily available.

An additional advantage of simulation as against alternative mathematical methods is the degree to which local influences of the data are taken into account. Where the characteristics of the deposit may depart locally from the assumed histogram or variogram model, the conditioning data actually override the model and produce a much more realistic local representation. Similarly, local geological features can be built in with much greater flexibility than in a more mathematical type of model.

LIMITATIONS AND CHALLENGES

The implementation study described in this paper has obviously shown up a number of limitations as well as the expected benefits of the use of conditional simulation, and much remains to be done to ensure a satisfactory integration of the technique with existing reserve estimation and mine planning technology.

The first problem, as discussed earlier, is one of credibility of the approach and difficulty in definitive validation. This is repeated here, since it is considered fundamental to the acceptance and successful implementation of any technology in a practical environment. While other approaches, or even alternative simulation algorithms to those adopted here, may be more aesthetically pleasing and mathematically rigorous, they can only be considered once they have a proven track record and have achieved a certain level of credibility in live application. This limitation means that the technology used may not always be optimal, but that the engineering adage of achieving 80% of the result for 20% of the effort is continually applicable.

The second challenge for effective implementation of conditional simulation is the integration of the results with existing techniques. The classical approach to mine planning and optimisation does not allow easily for concepts such as confidence or probability, and mine planning philosophies will need to evolve to make best use of this information. As discussed earlier, one of the main reasons for adopting a simulation approach rather than more mathematically-based techniques was to ease this evolution, but longer term developments might include the use of more efficient techniques than the computationally intensive simulation approach.

Lastly, the provision of additional dimensions of information leads in itself to some difficulties. As in all simulation applications, conditional simulation can produce large volumes of information very rapidly, and the degree to which this information is put to effective use is critical. There is always the tendency to accept results from such a method as being absolutely definitive, and to expect the approach to provide the answer to all problems. For conditional simulation to have sensible and effective implementation, a careful balance is required between the lack of credibility mentioned earlier, and the exaggerated hype that has seen the discrediting of other geostatistical methodology.

REFERENCES

Alabert, F. G. (1987) "The practice of fast conditional simulations through the LU decomposition of the covariance matrix", Math. Geol., v. 19, n. 5, pp. 369-387.

Boulanger, F. (1988) "Géostatistique et processus autorégressifs: une nouvelle méthode de modélisation" in M. Armstrong (ed), Geostatistics, Kluwer Academic Publishers, Dordrecht, pp. 259-272

Deutsch, C. V. and Journel, A. G. (1992) "GSLIB: Geostatistical Software Library and User's Guide", Oxford University Press, 336p.

Farmer, C. L. (1989) "The mathematical generation of reservoir geology", Paper presented at the Joint IMA/SPE European Conference on The Mathematics of Oil Recovery, Robinson College, Cambridge University, 25-27 July 1989.

Gómez-Hernández, J. J. and Srivastava, R. M. (1990) "ISIM3D: A three dimensional multiple indicator conditional simulation program", Computers and Geosciences, v. 16, n. 4, pp. 395-440.

Isaaks, E. H. (1990) "The application of Monte Carlo methods to the analysis of spatially correlated data", Ph.D. thesis, Stanford University, 213 p.

Journel, A. G. (1974) "Geostatistics for conditional simulation of orebodies", Econ. Geol., v. 69, pp 673-680.

Lantuéjoul, Ch. (1992) "Substitution Random Functions" in A. Soares (ed), Geostatistics Tróia '92, Kluwer Academic Publishers, Dordrecht, pp. 37-48.

Matheron, G. (1978) "Estimer et Choisir", Les Cahiers du Centre de Morphologie Mathématique de Fontainebleau

Nowak, M., Srivastava, R. and Sinclair, A. (1992) "Conditional Simulation: a mine planning tool for a small gold deposit" in A. Soares (ed), Geostatistics Tróia '92, Kluwer Academic Publishers, Dordrecht, pp. 977-988.

Srivastava, R. M., Hartzell, D. R. and Davis, B. M. (1992) "Enhanced metal recovery through improved grade control" in Proceedings of 23rd APCOM Symposium

Verly, G. (1992) "Sequential Gaussian co-simulation: a simulation method integrating several types of information" in A. Soares (ed), Geostatistics Tróia '92, Kluwer Academic Publishers, Dordrecht, pp. 543-554

Random media and lattice gas simulations

R. Brémond*, D. Jeulin**

* Centre de Morphologie Mathématique
** Centre de Géostatistique
Ecole Nationale Supérieure des Mines de Paris
35 rue Saint Honoré 77305 Fontainebleau

Abstract. Some possibilities of simulations using lattice gas models are illustrated. Starting from a lattice, the model, initially developped in statistical physics[5], generates random walks of a population of particles in mechanical interactions, respecting basic physical conservation laws (mass and momentum). The behaviour of this model reproduces viscous flows obeying the Navier-Stokes equations. When considering different types of particles, it is possible to introduce rules of interaction that mimic processes, based on transport phenomena, which generate random media on a lattice. This physical approach of simulations is illustrated by constructing random aggregates obtained by diffusion, nucleation and growth, under various hydrodynamical conditions.

Introduction. Many structures are the result of complex interactions between various components. A simple way to simulate these interactions is to build a physical model such as the lattice gas proposed in statistical physics [5]: the motion of particles on a lattice is constrained by basic conservation laws, such as the conservation of mass and momentum. This enables us to simulate the flow of a fluid obeying the Navier Stokes equations [4], and can be used to estimate the permeability or the coefficient of diffusion of porous media [12][7]. Starting with the lattice gas model, more complex situations are simulated by introducting new rules: two fluid flows accounting for surface tension [13]; liquid-gas mixtures to simulate the evaporation process [1], chemical reactions, phase separation, or deposition... In the present paper, we propose lattice gas models containing a suspension, to simulate aggregation processes. The growth of aggregates occurs in many fields, such as in the solidification from a melt including various surface tension effects, the filtration of a suspension by a porous medium, or the deposition of sediments. This is illustrated by the simulations presented in the next parts.

1 The lattice gas model and aggregation

The lattice gas model is a recent method to simulate hydrodynamics [6]. It is a discrete version that is free of inherent defects of classical numerical methods (boundary conditions, round-off errors), and that is easily implemented on general purpose computers.

M. Armstrong and P. A. Dowd (eds.), Geostatistical Simulations, 89–105.

1.1 Principles

Consider a periodic lattice, on a hexagonal graph. Each vertice of the graph is connected to its six nearest neighbours. It can contain between 0 and 6 particles with a mass and a velocity one (but at most one particle per velocity, corresponding to the six velocities of the grid). Every time step, the location of particles is changed according to the following local operations: a propagation of particles to the nearest neighbour in the direction of their velocity (Fig. 1.b), and a collision redistributing the velocities as shown in Fig. 1.c.

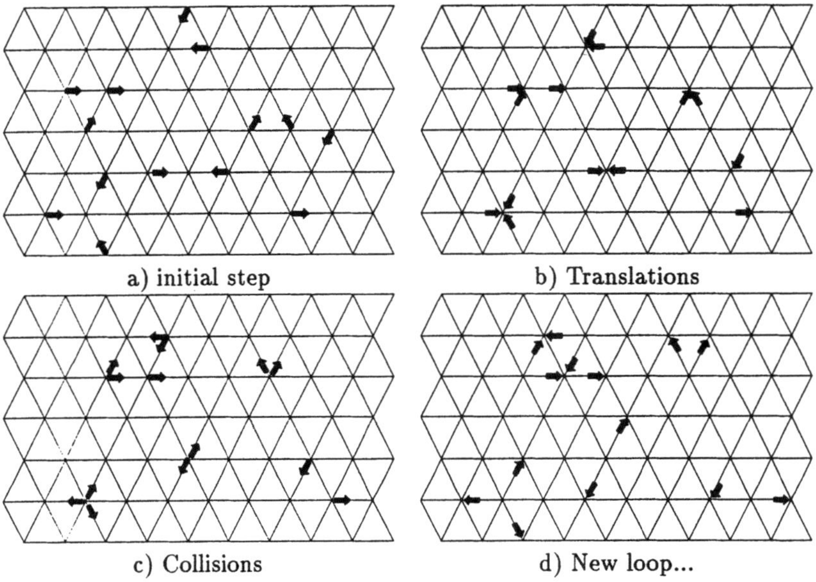

Figure 1 Lattice gas on a hexagonal grid, elementary step.

Collisions. The collisions are local operations that preserve the total mass and momentum (and consequently the kinetics energy) on each node. Fig. 1.c shows two configurations changed according to this transformation. In [4], it was shown that this model, named FHP, follows the Naviers-Stokes equation on a macroscopic scale. Therefore, the simulation of complex flows is available from lattice gases. The rules used in this study are the FHP IV rules given in Fig. 2.

Boundary conditions. The physical boundary conditions are easily handled. Each vertex of the graph can be referenced as a pore, or as an obstacle. In the latter case, the collision is replaced by bounce-back conditions. This is equivalent to a no-slip condition on the boundary. On the edges of the field, periodic conditions (particles leaving the field on one side are reinjected with the same velocity on the opposite side), as well as non periodic conditions (with random injection of particles) are implemented.

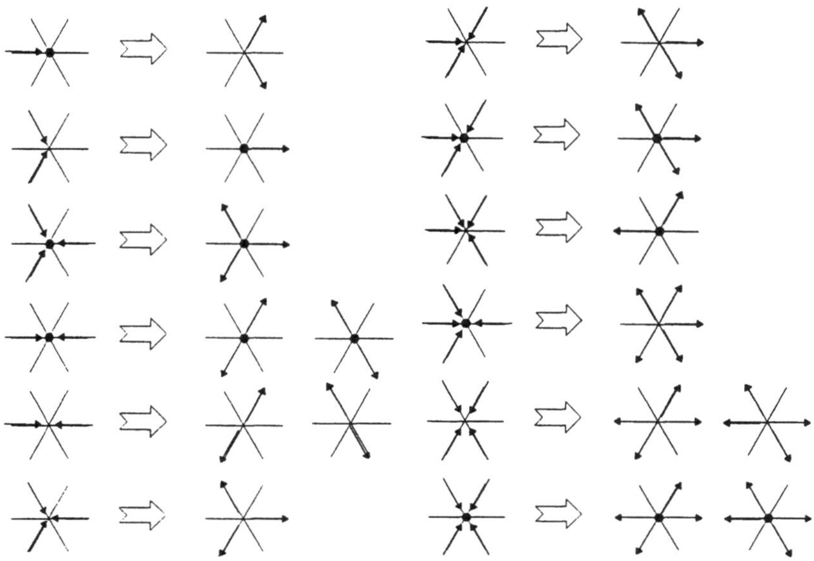

Figure 2 FHP IV model, collision rules.

Marking particles. To simulate a mixture of fluid and suspensions, we use marks (F and S). The collisions are processed as follows: the standard rules are applied on each node, whatever the marks; the marks are redistributed at random after the collisions. The behavior of the two types of particles, F and S, differs only during the aggregation process.

1.2 Initial conditions.

The simulation can be considered as an iterative process converging to a velocity map which is a solution of the hydrodynamical problem with its boundary conditions. Since in general this solution is not known, we start from independant and uniform velocities. After stabilization of the resulting field, the aggregation process can start.

1.3 Aggregation.

Various rules of aggregation can be introduced. We operate on a field containing pores and obstacles (including aggregates). When a particle becomes the nearest neighbour of obstacles, it is candidate to aggregate. The following rules are applied: a probability of aggregation, p^+ is introduced for each particle (which is bounced back with the probability $1 - p^+$). Conditional rules can be added: the particles are allowed to aggregate as long as they are moving towards an obstacle (Fig. 3.b), when they move along the obstacle (Fig. 3.c), or even whenever they are distant one pixel from the obstacle (Fig. 3.d). This is made by testing the presence of an obstacle in the neighbourhood inside a cone of 0, 60 or 180 degrees (Fig. 4).

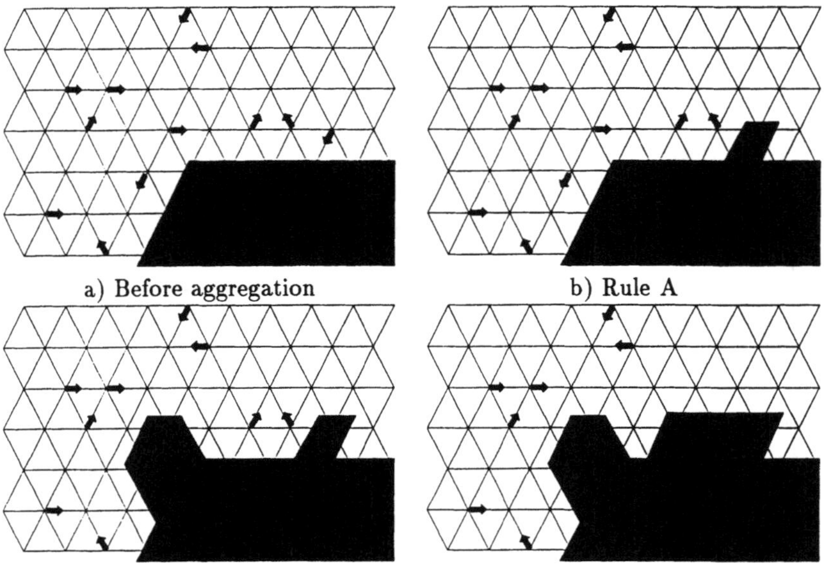

a) Before aggregation b) Rule A

c) Rule B d) Aggregation in the neighbourhood

Figure 3 Different rules of aggregation.

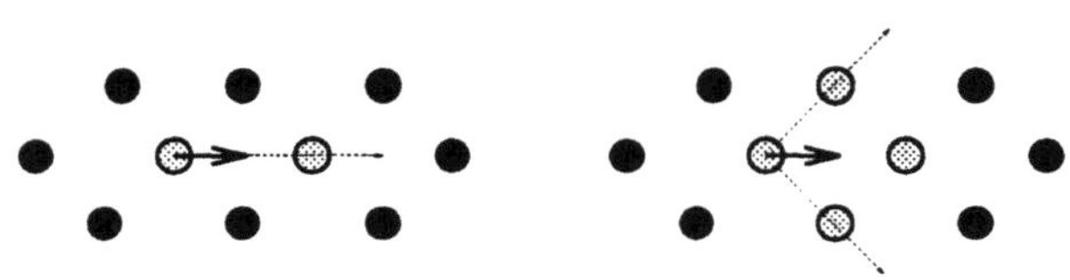

Figure 4 Directionnal neighbourhoods.

Additional conditions concern the number of neighbours of a suspension located in an aggregate. In Fig. 5 particle a possesses two neighbours in the black aggregate, particle b has 3 neighbours, and particle c a single neighbour. Rules depending on the number of aggregated neighbour results in different geometries. Other rules are physical, such as the local pressure (proportional to the particle density), which can stop the aggregation if it is too high. Particles in the aggregate can leave the aggregate for the fluid with a probability p^{-} in order to simulate a disintegration. In what follows, simulations obtained from three rules will be compared: rule A (Fig. 3.b), rule B (Fig. 3.c), and rule C, similar to rule B, but where the suspensions can aggregate if at least 2 of the 3 considered neighbours are already aggregated. These conditions mimic some surface tension effects which occurs ins the solidification process.

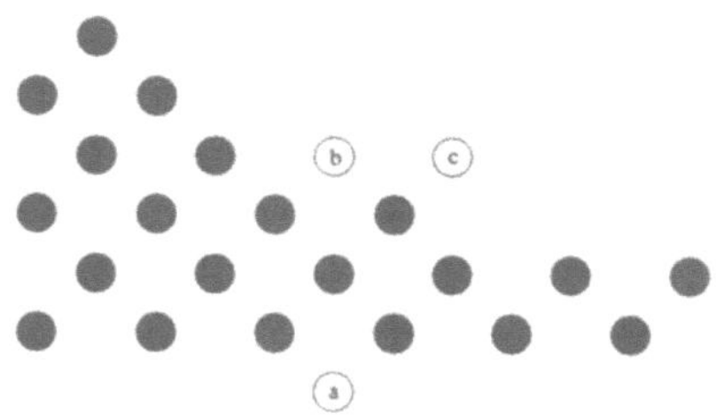

Figure 5 Neighbourhood configuration.

2 Simulations of random aggregates

As an illustration of the model defined in the previous section, some simulations are provided and interpreted in physical terms. The following situations are presented: growth of aggregates under various boundary conditions, growth of populations of aggregates, deposition of suspensions. Finally, the morphology of the simulated structures is characterized by measurements.

2.1 Aggregates from point seeds

To visualize the effect of the three rules A, B, and C, simulations were made on a periodic system with size 200×200. Starting with a solid seed in the middle of the field, the aggregate grows by diffusion for a zero average velocity, as in the case of the DLA (Diffusion Limited Aggregation) model [9][15]. However in the present simulations, the change in the local density of particles and the perturbation of the velocity field by the aggregate are automatically satisfied. The initial densities of fluid and suspensions are 2.1 and 0.21 particles/pixel.

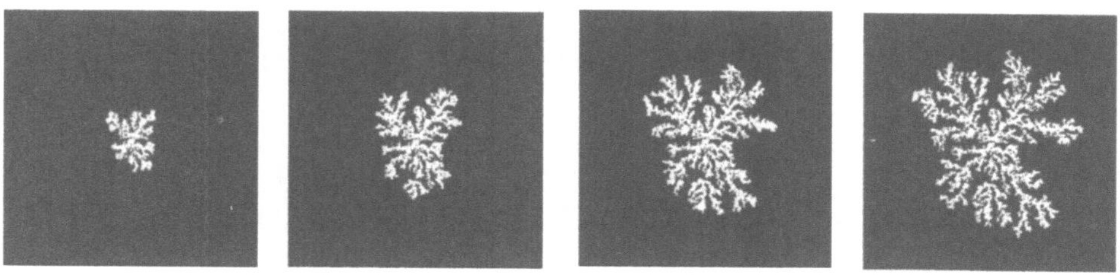

a) $t = 500$ b) $t = 1000$ c) $t = 1500$ d) $t = 2000$

Figure 6 Growth of an aggregate from a central seed, rule A.

In Fig. 6, the growth of the aggregate, obtained by rule A, is shown for increasing steps of time. By adding the disintegration process, it is shown in Fig. 7 that more and more compact objects are obtained when increasing p^-. A comparison of the results of rules A, B, and C without disintegration and with $p^- = 0.005$ is given in Fig. 8 and Fig. 9. Qualitatively, there is a progression towards more and more compact aggregates obtained

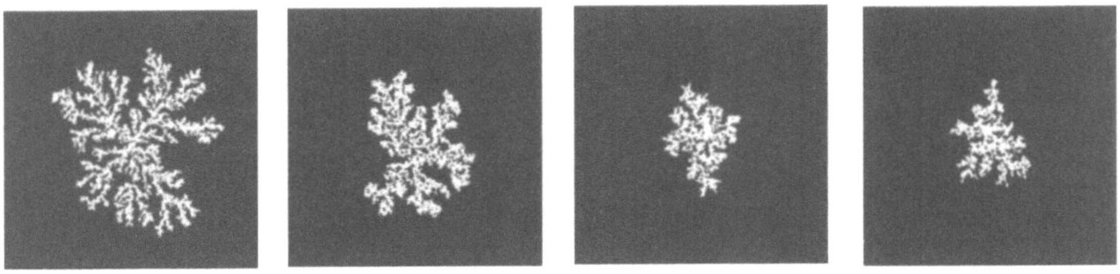

a) $p^- = 0.0$ b) $p^- = 0.002$ c) $p^- = 0.004$ d) $p^- = 0.008$

Figure 7 Change of the morphology of the aggregate with p^-, rule A, $t = 2000$.

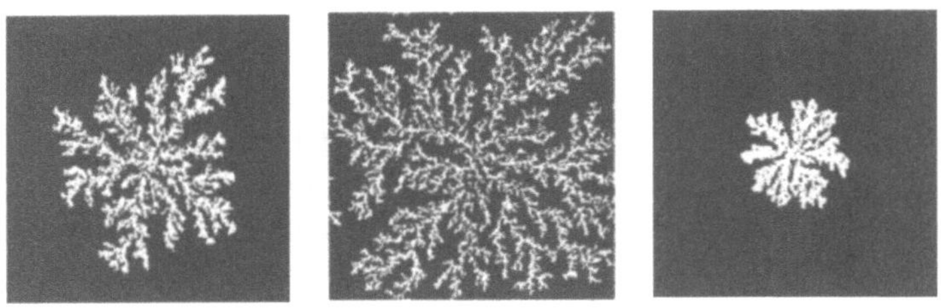

A B C

Figure 8 Comparison of the different rules of aggregation with $p^- = 0.0$, $t = 2000$.

by rules B, A, C. The last one can be compared to dendrites grown from a melt in the presence of a surface tension between the solid and the liquid. The packing effects resulting from the p^- coefficient weaken the differences of morphology between the different rules (Fig. 9).

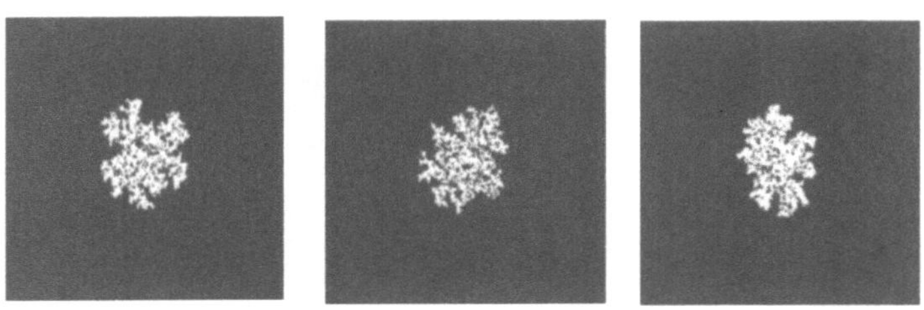

A B C

Figure 9 Comparison of the different rules of aggregation with $p^- = 0.01$, $t = 2000$.

Other boundary conditions introducing a non zero velocity field are interesting. A horizontal shear is introduced by forcing the flow in two opposite directions on the upper and lower boundary (Fig. 10). The resulting aggregate is no longer isotropic. Vertex conditions are obtained by forcing the flow around the edges of the largest hexagon included inside the field (Fig. 11). The aggregates are still isotropic, but their branches are parts of spirals starting from the initial seed.

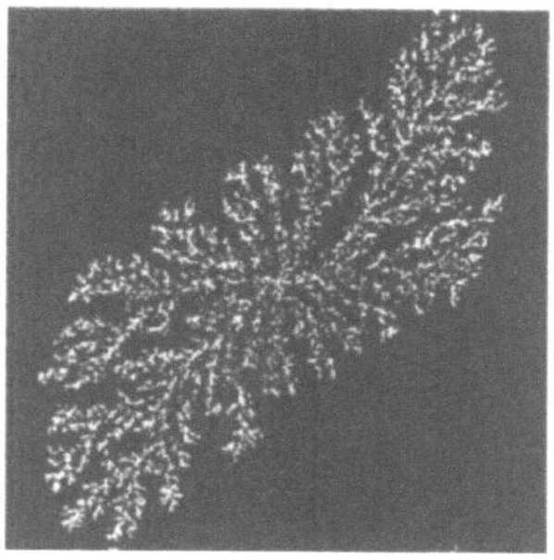

Figure 10 Aggregate obtained with shear boundary conditions, $t = 5000$.

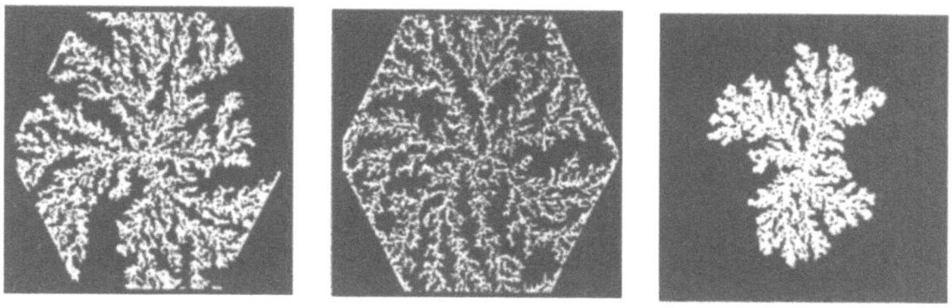

A B C

Figure 11 Aggregates obtained with vortex type boundary conditions. comparison of the rules A, B, C, $t = 2000$.

2.2 Nucleation and Growth of Aggregates

By replacing the single seed of the previous section by a population of Poisson points, structures very similar to the dendritic solidification out of a melt are obtained. In Fig. 12, the intensity of the Poisson process is equal to 0.001, representing an average of 40 seeds on 200×200 images. The aggregation is made by diffusion, as previously, but there is now competition between the seeds for trapping the particles. In addition, the aggregates are allowed to coalesce when they enter into contact, as seen for the case of rule B. By construction the structure is periodic vertically and horizontally.

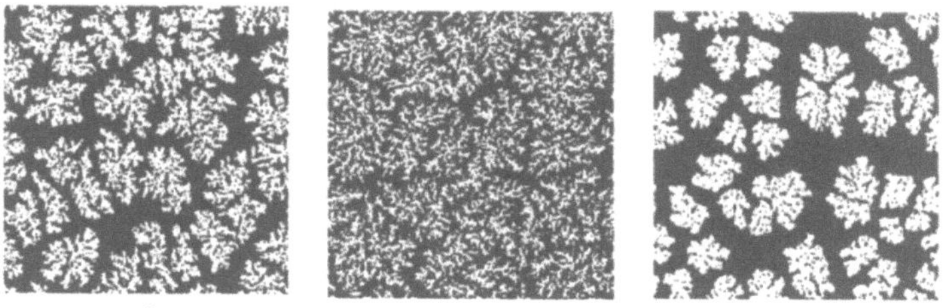

A B C

Figure 12 Growth from Poisson seeds for rules A, B, C,$t = 2000$.

The same process was simulated with a continuous introduction of random seeds producing the nucleation of aggregates. In Fig. 13, the simulation is observed after 2000 iterations, with a new seed every 100 time steps. The nucleation produces a dispersion in the age of aggregates resulting in a distribution of sizes.

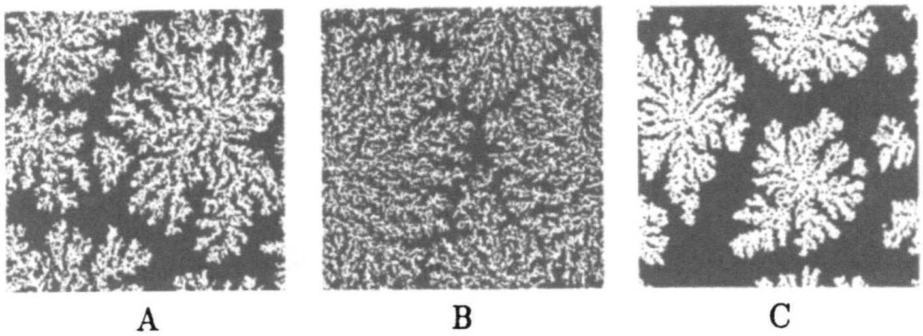

Figure 13 Germination and growth from Poisson seeds for rules A, B, C, $t = 2000$.

2.3 Deposition

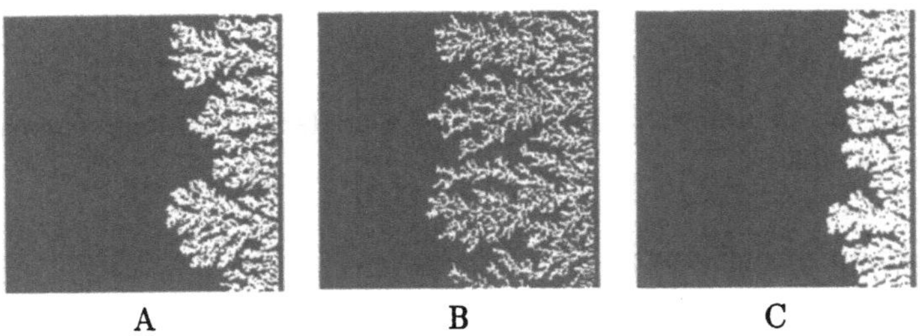

Figure 14 Aggregates obtained by deposition, $t = 5000$.

The deposition of particles occurs in many situations, including during the process of sedimentation operating on geological scales. In a set of simulations, the deposition of particles in suspension, subject to a vertical force, such as gravity, is obtained as follows: a medium is open and periodic on its vertical boundaries, and closed on its horizontal boundaries. Particles in the suspension are moved downwards by a force, implemented by the collision rules, and can stick on the lower boundary. Therefore, there is a combination of diffusion and convection. Examples are given in Fig. 14 (where the gravity is horizontal) for the three aggregation rules. They are very similar to the previous results obtained for aggregates generated from point seeds and can be viewed as a dendritic solidification process on a cold plate. After introduction of a probability of disintegration, the gravity field produces a packing of the structure (Fig. 15).

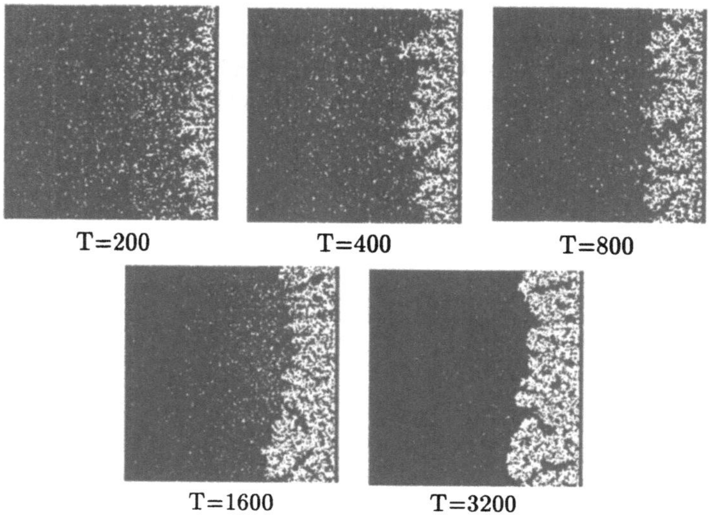

Figure 15 Sequence of packing of an aggregate ($p^{-}= 0.005$; rules B); the moving suspensions are displayed.

3 Measurements on Aggregates

A complete characterization of the simulated random media, X, would require the theoretical calculation (or the practical estimation) of their Choquet capacity [8]:

$$T(K) = P\{K \cap X \neq \emptyset\} \tag{1}$$

In equation 1, K is a compact set. Here we operate on countable sets of points on a lattice, so that X is characterized by its spatial law, which is not known from theoretical calculation. Some measurements are used to compare the morphology of aggregates, such as the area and perimeter, the " fractal dimension ", the number of connected components, and the distribution of areas. In addition, the change of some morphological properties with time is useful to study the dynamics of the process. We conclude with the covariance of aggregates.

	Perimeter	**Area**	**P/A**
Rule A	4590	5120	0.896
Rule B	8360	8370	0.998
Rule C	1550	2750	0.565

Table 1 Area and perimeter of aggregates obtained for rules A, B, C.

The basic measurements, and the ratio Perimeter/Area, estimated as an average over 10 realizations per rule of aggregation, after 2000 iterations, are given in Table 1. The ratio P/A decreases in the order B, A, C, reflecting more and more compact aggregates. The variations of areas for the different rules are due to the different growth rates, as measured below.

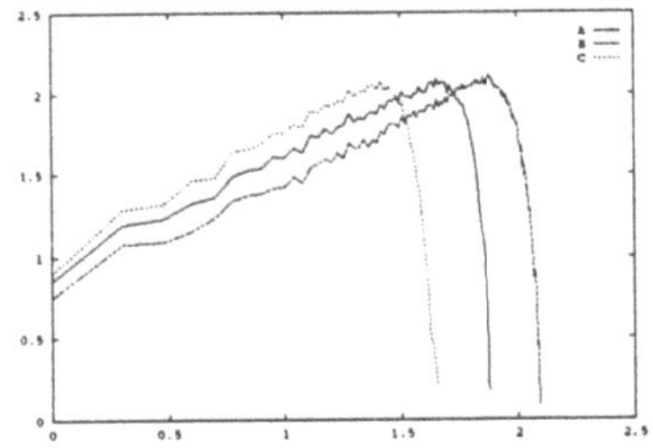

Figure 16 Log($S(r)$) versus Log(r).

The " fractal dimension " α of aggregates is estimated from equation

$$S(r) = Ar^{\alpha-1}dr \quad (2)$$

where S(r) is the area of the aggregate included inside a crown of radius r and of thickness dr, centered on the seed of the aggregate. The average over 10 aggregates of log(S(r)) is reported in Fig. 16 as a function of log(r), and the estimated fractal dimension, obtained from the linear part of the curve, is reported in Table 2 with its fluctuations when considering different portions of the curve. To compare the results of the different rules, it is wise to keep a common linear part of the three curves in Fig. 16, namely the domain corresponding to $r < 20$. The same measurement was made for the aggregates built with vortex boundary conditions.

	Range	r < 20
Rule A	1.72 - 1.74	1.736
Rule B	1.64 - 1.71	1.672
Rule C	1.80 - 1.82	1.813
Rule B Vortex	1.70 - 1.76	1.703

Table 2 Fractal dimension of aggregates.

More important fluctuations in α are observed with rule B under any boundary conditions. It increases in the order B, A, C, just as for the ratio A/P, reflecting a decrease in the roughness of aggregates. For comparison, the parameter α obtained for the DLA model built with random walks on a square grid is equal to 1.715 [10]. It is very close from the result of the present simulations with rule A. However care must be taken to draw conclusions: The range of scales used to estimate α is limited to one order of magnitude (as for simulations on random aggregates by other methods). In addition, since a large amount of uncertainty is observed on such measurements, and the fractal dimension is only a part of the description of the roughness of aggregates on a small scale, it does not account for morphological changes induced by a vorticity or a shear boundary condition.

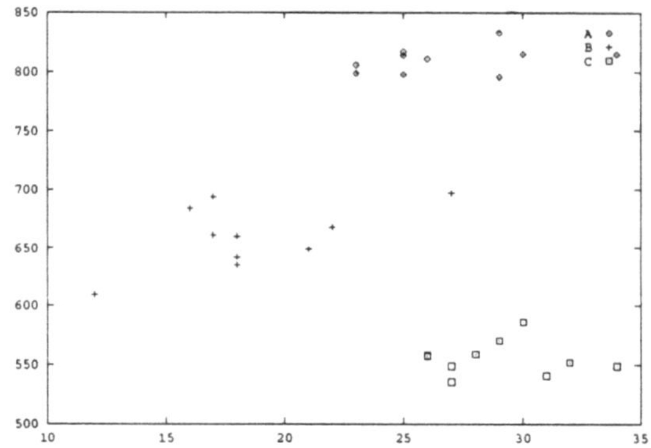

Figure 17 Number of connected components of the background, versus the number of connected components of aggregates.

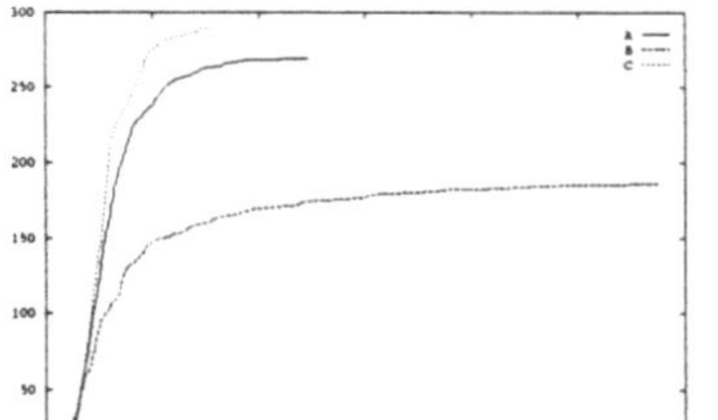

a) Poisson seeds

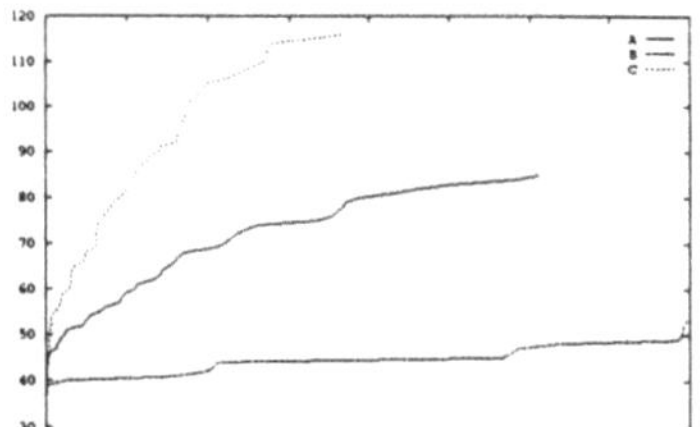

b) Germination and growth

Figure 18 Cumulative distribution of areas (in numbers).

Connectivity. The connectivity of the population of aggregates grown from Poisson seeds is characterized by the numbers of connected components of the aggregates and of the background for each simulation, as reported on Fig. 17. The counting respects the periodicity. An aggregate growing across a boundary is counted once, so that the number of connected components in a period is measured. The three geometries corresponding to rules A, B, C, differ (Fig. 17). Rule A produces the largest number of components for the aggregates and the background; rule C gives a similar number of aggregates as rule A, but since they are less porous, the number of connected components of the background is much lower; rule B leads to a lower number of objects, as a result of the coalescence of more ramified aggregates.

The distribution of areas, given as cumulative numbers in Fig. 18, reflects the smallest number of connected components for the nucleation and growth conditions, as an effect of the coalescence. The area of the largest connected component increases in the order C, A, B, as was already clear from Fig. 12 and 13.

Dynamic measurements were recorded as a function of time: the mass of aggregates, the length of the interface between aggregates and pores P, and the age of individual aggregates at the end of the simulation. In Fig. 19.a, the growth rate of a single aggregate (averaged over 10 simulations) first increases and then decreases when free suspensions become very sparse, as seen for rule B where the growth rate is the higher. In the presence of disintegration, the three types of aggregates have similar growth rates, with a drop in the mass when large parts of the aggregate are disconnected (Fig. 19.b).

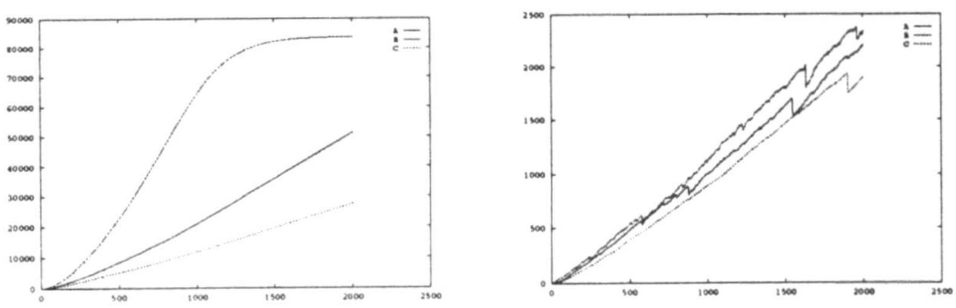

a) $p^- = 0$ b) $p^- = 0.005$

Figure 19 Area of the aggregates as a function of time.

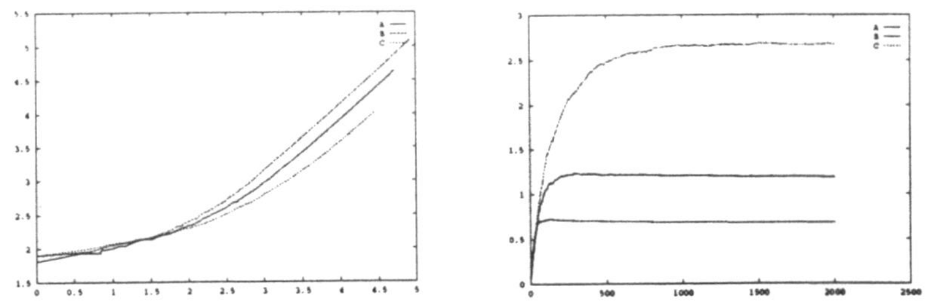

a) $Log(A)$ versus $Log(P)$ b) Ratio $\frac{A}{P}$

Figure 20 Relationship between the area A and the length of contact P.

The change of the length of contact with time is very similar. It is more instructive to examine the evolution of the ratio A/P with time. This ratio reaches a limit (the growth rate per unit of time and per pixel), which depends on the rule of aggregation (Fig. 20.b). In Fig. 20.a, it is clear that two rates of growth appear, with two steps power law dependences of the area as a function of the perimeter (between 0.21 and 0.28 at the beginning, and 1 at the end of the process). As noticed in [14] for a different model, the second step corresponds to the dendritic growth of aggregates.

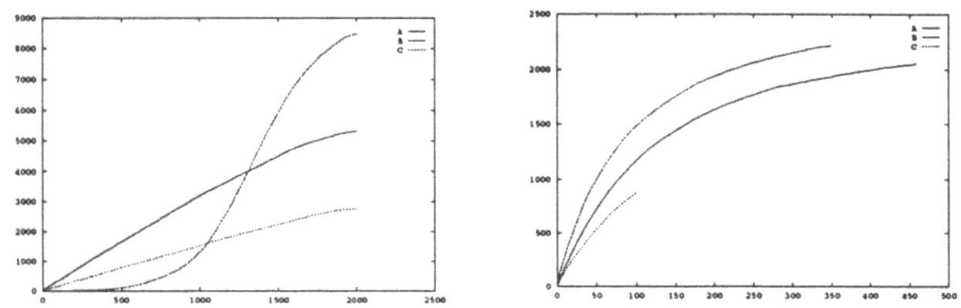

a) $p^- = 0.0$ b) $p^- = 0.01$

Figure 21 Cumulative distribution of the age of aggregates.

The distribution of the ages of particles in aggregates at the end of the simulation with point seeds is given in Fig. 21. In Fig. 21a, ten simulations are considered for each aggregation rule. For rule B, there are very few recent particles, since the growth was finished before the end of the simulation. In Fig. 21.b, the particles in aggregates are younger as a result of the disintegration.

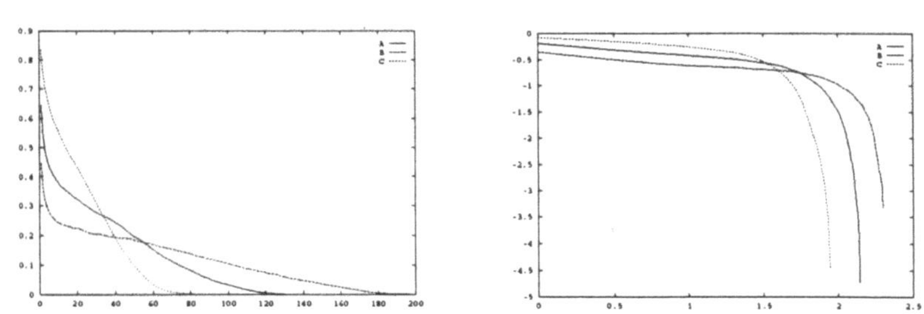

a) Covariogram b) Log-Log representation

Figure 22 Geometric covariogram of individual aggregates grown from a single seed, for rules A, B, C, with $p^- = 0.0$.

Second order statistics. The geometric covariograms of aggregates obtained from a point seed are given in Fig. 22. They have a power law behaviour for short distances, with exponents increasing in the order B, A, C, as for the fractal dimension. The covariances of the other structures are given in Fig. 23. They reflect the sizes of the structures.

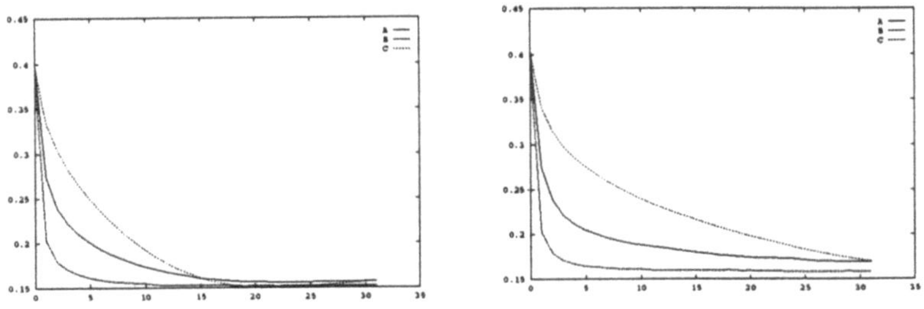

a) Poisson seeds b) Germination and growth

Figure 23 Covariances C(h) of the population of aggregates.

4 Aggregation and filtration in porous media

In this section, the lattice gas simulation is used to generate microstructures from the flow of a suspension in a porous medium. It is obtained in two-dimensions by erosion of Voronoï polygons, built as zones of influence of Poisson points. The medium is extended to infinity by periodicity. Therefore the fluid can circulate in a network of interconnected channels.

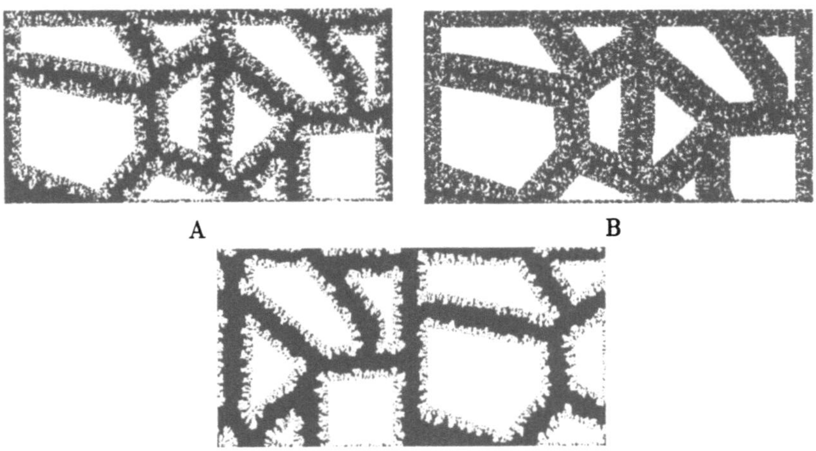

A B

C

Figure 24 Aggregates obtained by diffusion in porous media, $t = 2000$.

Diffusion and aggregation in porous media. Using a zero average velocity, particles move by diffusion before their aggregation on the boundaries of channels (Fig. 24). This is similar to a secondary solidification after partial melting of rough particles. From the 256×512 images, the " covering ratio " (proportion of the boundary of grains covered by aggregated suspensions) is estimated, and reported in Table 3, where it appears that this ratio increases when the rules of aggregation produce more compact aggregates.

Rule A	Rule B	Rule C
0.686	0.410	0.835

Table 3 Covering ratio of the grain boundaries.

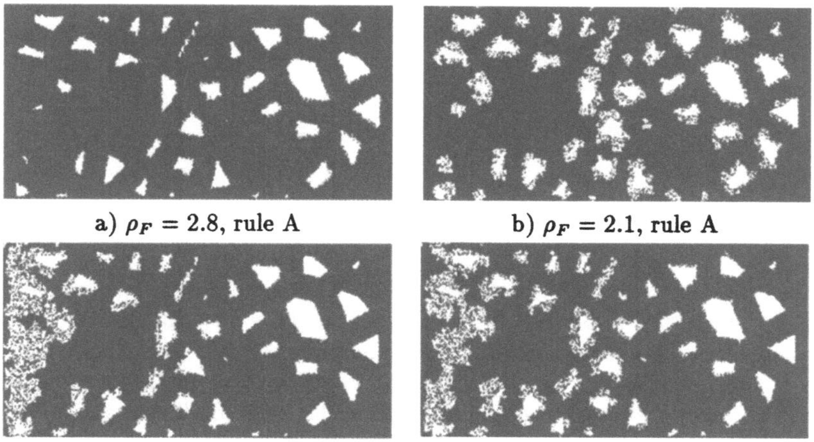

a) $\rho_F = 2.8$, rule A b) $\rho_F = 2.1$, rule A

c) $\rho_F = 1.4$, rule A d) $\rho_F = 2.1$, rule B

Figure 25 Aggregates grown in flows in porous media, $t = 5000$.

Filtration in porous media. Other simulations can reproduce the flow of the suspension in a porous medium. The aggregation on pore boundaries, as well as the percolation of aggregates is similar to a filtration process and to the clogging of the filter, as studied elsewhere [2]. For the four simulations shown in Fig. 25 after 5000 iterations, the medium is vertically periodic. The fluid is introduced on the left with the density ρ_F. This creates a pressure gradient carrying the fluid from left to right. The right wall of the medium is open, so that particles of the lattice gas can enter or leave the field (with size 256×512). The intensity of suspensions in the entering fluid is $\rho_S = 0.28$ suspensions/pixel. A new rule of aggregation is introduced in addition to the previous ones: a suspension is able to aggregate when there is no fluid particle (zero pressure in the fluid) on the same pixel. The coefficient of disintegration is set to $p^- = 0.01$. For the larger fluid density $\rho_F = 2.8$ (Fig. 25.a), there is almost no aggregation, while for $\rho_F = 1.4$ (Fig. 25.c), the porous medium behaves like a granular filter that is progressively blocked on the entering side. It can be seen from Fig. 25.b and d that aggregates obtained by rule B can percolate, while the fluid can still flow for rule A. In these simulations, the filtration process can be studied for various geometries, using correct hydrodynamics conditions. For instance, the overall flux of fluid during the filtration can be recorded, and compared to experiments. On Fig.

26, the slowing down of the instantaneous flux is induced by the clogging, which leaves narrower paths for the fluid.

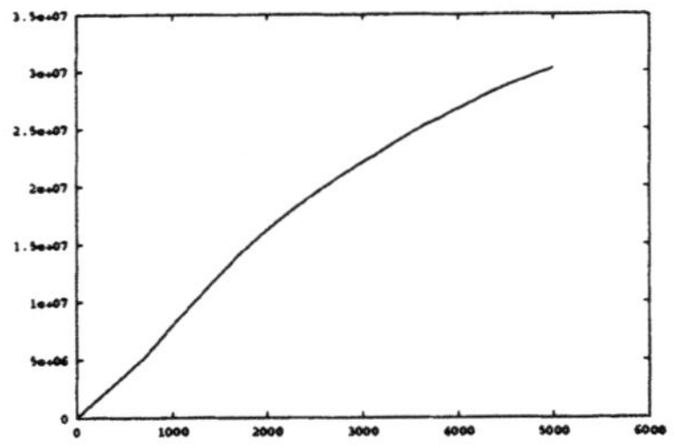

Figure 26 Cumulative flux, $\rho_F = 2.1$, rule A.

5 Conclusion

The flexibility of the lattice gas model enables us to simulate the generation of media regulated by hydrodynamical rules. This was illustrated here by the formation of random aggregates under various conditions. The strong coupling between the velocity field and the geometry is satsfied with this simulations, as opposite to other methods. Other points of interest can be introduced, such as chemical reactions between species [3], and 3-dimensional simulations [11]. In our approach, the scale of interest was microscopic. Other scales of time and of space can be considered as well, to simulate diagenisis processes. Finally, a further step will include conditional simulations.

References

[1] C. Appert, D. Rothman, and S. Zaleski. *A liquid-gas model on a lattice.* G.D. Doolen, Physica D, pp. 85-96, 1990.

[2] R. Brémond, D. Jeulin, C. Dathy, and M. Abouaf. *Simulation par gaz sur réseau de la filtration de la fonte.* Mem. Sci. Rev. Métallurgique, n°9, p.534, September 1992.

[3] D. Dab and J.P. Boon. *Cellular automata approach to reaction-diffusion systems.* Cellular automata and modeling of complex physical systems, pp. 257-273, 1989.

[4] U. Frish, D. d'Humières, B. Hasslacher, P. Lallemand, Y Pomeau, and J.P. Rivet. *Lattice gas hydrodynamics in 2 and 3 dimensions.* Complex System 1, pp.75-136, 1987.

[5] U. Frish, B. Hasslacher, and Y. Pomeau. *Lattice-gas automata for the Navier-Stokes equation.* Phys. Rev. Lett, pp. 11-18, 1986.

[6] J. Hardy, Y. Pomeau, and O. de Pazzis. *Molecular dynamics of a classical lattice gas : transport properties and time correlation functions.* Phys. Rev. A, 13, pp. 1949-1961, 1976.

[7] D. Jeulin. *Flow and diffusion in random porous media.* Numerical methods for the simulation of multi-phase and complex flow, Springer-Verlag, pp. 106-123, 1992.

[8] G. Matheron. *Random sets and integral geometry.* Wiley, 1975.

[9] P. Meakin. *Computer simulation of growth and aggregation processes.* On growth and forms, fractal and non-fractal patterns in physics. Martinus Nijhoff publishers, pp.111-135, 1986.

[10] P. Meakin and L.M. Sanders. Phys. Rev. Lett, vol 54, p. 2053, 1985.

[11] J.P. Rivet, M. Henon, U. Frisch, and D. d'Humières. *Simulating fully 3D external flow by lattice gas methods.* Discrete kinematic theory, lattice gas dynamics and foundations of hydrodynamics, World Scientific, pp. 276-285, 1989.

[12] D.H. Rothman. *Cellular automaton fluids: a model for flow in porous media.* Geophysics 53, pp.509-518, 1988.

[13] D.H. Rothman and J.M. Keller. *Immiscible cellular automaton fluids.* Lattice gas methods for PDE, Addison-Wesley, pp.275-282, 1989.

[14] T. Vicsek. *Formation of solidification patterns in aggregation models.* Phys. Rev. A, vol 32, p.3084, 1985.

[15] T.A. Witten and L.M. Sanders. Phys. Rev. Lett, vol 47, p. 1400, 1981.

DISCUSSION FOLLOWING SESSION NO 3

Chairman: Margaret Armstrong

Papers: Ravenscroft; Bremond & Jeulin

After both paper's were finished, the chairman asked if there were any questions on Jeulin's paper.

Séguret: My question has 3 parts to it:
1) How do the results depend on the initial germ?
2) What criterion is used to stop the iterations?
3) I agree concerning the Navier–Stokes property but do these simulations
respect any special statistics such as the correlation function. Is this possible?

Jeulin: In reply to the first question, clearly if I have an aggregate like this (he points to a transparency) then I know that this point was the origin of the coordinates. In this case we just wanted to simulate a random set (a random compact set at the origin). Then if you use a Poisson distribution of points you can simulate nucleation at Poisson points and then growth. You can change the process as well, so you have considerable freedom in what you do.

Second point, concerning the number of iterations, we start with a given suspension density and during the time the suspension lands on the aggregates and so the suspension concentration outside the aggregate drops. Consequently the process stops when there is no more in the suspension. If you continue to introduce suspension, the process stops when the whole region is full, except for parts that can no longer be reached due to the connectivity. In the present case, there are a limited number of particles so the process stops when these have all be used up. Last point concerning the statistics, we do not know much from theory. For example, the geometrical covariance is unknown for this kind of model. But these can be measured at different times on the simulations. For example here we see that it is a power law behaviour at the origin. We also measured the covariance for the population of aggregates. The results are given in the paper. The statistics can be measured on the

M. Armstrong and P. A. Dowd (eds.), Geostatistical Simulations, 106–109.

simulations and not given by a theoretical calculation. What is interesting here is to study the dynamics of the system; for example, the evolution of the area of the aggregates with time. Here it is not uniform. And so you can study the physics of the growth process that is behind your model. I repeat this has to be done experimentally from the simulations. It seems to be hard to calculate the geometrical covariance or the Choquet capacity theoretically.

One thing that, in my opinion, is important for the quality of the simulations is to look at the simulations and compare them with reality. There is a lot of information in their shape. You can compare them visually with the real objects that they are supposed to model, and decide whether or not they look alike.

Armstrong: This question is for Peter Ravenscroft. This morning in Carol Gotway's talk we were looking at 4 particular loss functions all related to environmental science. Coming back to mining, what types of loss functions would you suggest using to compare simulations with reality?

Ravenscroft: Initially, one should check the things that one had set out to reproduce (the covariance function and the histogram). Now for loss functions, I would be interested in the distributional aspects such as the quantity above a given cut off grade. You should produce something which matches what you want – which matches reality, in the range that is of practical importance. I think you should concentrate on the distributional aspects or cutoff criteria.

Kleingeld: Peter, if you do not have enough data to get a good estimate, why should your simulation give a better impression of the deposit? Do you believe you can?

Ravenscroft: No, the converse is probably true. If you cannot get a decent linear estimate, you will probably get a bad simulation. If you are doing multiple simulations, then in that case your simulations turn out to be very spread out. One of the things that I should have said earlier, is that the power that I see in the conditional simulation approach is that you start out with a global parametric model (in terms of the variogram and the histogram) and if you have areas of the deposit where your data show departures from that model, then your conditional simulations will depart from the model as well. So the simulation will actually turn away from the model and match the data, where there is enough data.

Kleingeld: So we come back to the concept of having enough data because one problem that we must face, is of getting a given variogram and histogram from different underlying models. So the variogram/histogram criterion may not be enough to judge conditional simulations. Moreover in conditional simulations, the data force the simulation to follow. So if there is an isolated high data value, then the surrounding simulated values will not necessarily pick it up.

Ravenscroft: You are quite right. Your data will bind the simulation at that point, and that value could be missed.

Kleingeld: One last question. If you worked your grade tonnage curve out from drillhole or blast hole data, after allowing for the support effect and the information effect, would it differ very much from the one obtained from simulation?

Ravenscroft: Again, if the model used to generate the support and information effects is applicable, then you should get the right result. In our experience, even that is a global generalization of what might be happening locally. So for example you might have a deposit with a very strong continuity of high grades or low grades (or the converse) and you do not see that in the global model but the simulation model could bring it out.

Kleingeld: I agree totally with you, that the simulation could give a better local grade–tonnage curve for planning purposes but as far as the global ore body is concerned, I do not think that it is necessary.

Ravenscroft: You are quite right. But the question that I ask you is "of what use is a global grade–tonnage curve?".

Kleingeld: It is used to estimate the overall potential of an orebody, but to realize that potential one has to put in more drillholes to identify the first five years' mining correctly.

Ravenscroft: I have not got down to that level of detail yet. I am talking about scheduling an open pit in terms of annual phases or whatever. In early years you should be mining a high grade ore zone with its particular grade–tonnage curve; in later years you could be assessing lower grade ore with a different grade–tonnage distribution. I agree that the mathematical approach via a model is much quicker and much more satisfying, for a first result, but we also feel that you get more additional information from a simulation because your get a block model that can be used for conventional mine planning.

Kleingeld: Yes but the block model depends very much on the conditioning.

Armstrong: Earlier you said that accepting a simulation required an act of faith. Are you turning simulations into a religion now?

Ravenscroft: No but as I said if you set yourself the objective of only accepting a model that you can prove absolutely, then we would all still be back in the 18th

century. I think you have to accept the idea that you cannot validate it absolutely but that you have done all the right things in constructing model, then you have to take that step.

Bourgine: I have two questions for Dominique Jeulin. Firstly how do you set the initial conditions for the model:– the number of particles and their speeds and directions. Are the results very sensitive to these conditions? Secondly, you said that the particles have mass and velocity equal to 1.0. Is it possible to give a histogram of masses and velocities?

Jeulin: In answer to the first question, this is an iterative process. You can just describe this model, these simulations, as an iterative process that converges toward the solution of the problem (i.e. the velocity field in terms of the boundary conditions). If you start with the solution, in two iterations you have finished. In fact, what we do is to start from a priori information. With no prior information we start with zero velocity on average but with randomness. I mean velocities are set at random and independent from pixel to pixel. So it takes some iterations of the process before you get a solution. Getting a solution is not a trivial problem. We wait till this stabilizes before starting the process of validation. We have to be sure that we are working in a good velocity field. But after that, during all the iterative procedures, the velocity field is an input data. But we usually start with white noise for the velocities. We could start with something else and the result would be the same, but it might take more time to return the velocities.

Now concerning the other point, yes some people have tried to do this. Two problems arise. Having different velocities on the lattice means that the collision rules become extremely complicated. The advantage of using different velocities is that it represent gas particles with different energies. In the current model, the energy is constant because energy is 1/2 mv^2 and both m and v have been set at 1. If you let the energy vary, you can simulate thermal fields. Some people have done this. You can imagine the complexity that this leads to and the number of configurations that have to be considered. So it is preferable to work with a very simple model.

After thanking both speakers, the chairman closed the discussion.

Theory and Practice of Sequential Simulation

J. Jaime Gómez-Hernández and Eduardo F. Cassiraga
Departamento de Hidráulica, Universidad Politécnica
46071 Valencia, Spain.

Abstract

Sequential simulation is a powerful stochastic simulation technique the theory of which relies on the ability to determine, for a given multivariate model, the conditional probability of a single random variable given any number of conditioning values. Sequential simulation can be used with those multivariate models for which these conditional probabilities can be determined. In practice, it is not enough to know how to determine the conditional probabilities, the procedure must be feasible from an operational point of view. Because it is not so in most cases, some approximations to the conditional probability distribution function are used in the implementation of the technique. These approximations are shown not to have a large impact in the performance of the technique, at least for the case in which the underlying multivariate model is multiGaussian.

Introduction

Sequential simulation is, from a theoretical point of view, a very powerful simulation technique. Its theoretical basis is very simple and easy to understand and it can be accommodated to many different simulation problems, such as the simulation of a single variable, either continuous or categorical, or the joint simulation of multiple correlated variables.

Sequential simulation is, from a practical point of view, still a very powerful simulation technique. There are some implementation problems that must be solved to make the technique feasible and some limitations of which the user must be aware.

Although sequential simulation got into the geostatistical arena more than five years ago (Alabert, 1987), and although it has been applied in many studies since, it was last year when a surge of theoretical papers appeared on the subject (Deutsch and Journel, 1992; Gómez-Hernández and Journel, 1993; Verly, 1993; Omre *et al.*, 1993). This paper reviews the theory of sequential simulation and presents some of the practical problems encountered during its implementation.

Theory

Consider the joint distribution of a total of N random variables. The N random variables comprise K different attributes over the N' nodes of a grid, with $N \leq KN'$,

M. Armstrong and P. A. Dowd (eds.), Geostatistical Simulations, 111–124.

that is, there are nodes at which not all attributes are to be simulated. The generic notation Z_i, with $i \in \{1, \ldots, N\}$ is used to denote any such random variable no matter its location or its attribute-type.

Next consider the set of N_0 data of any type. The purpose of stochastic simulation is to generate realizations of the N random variables honoring the N_0 data, that is, to draw N values at random from the multivariate conditional probability distribution function (cpdf) $f(z_1, \ldots, z_N|(N_0))$, where $|(N_0)$ denotes conditioning on the N_0 data values.

Random drawing from the N-variate cpdf can be accomplished by sequential drawing from the N univariate cpdfs that result of iterative application of the definition of conditional probability to an ordering of the N variables:

$$\begin{aligned} f(z_1, \ldots, z_N)|(N_0)) &= f(z_N|(N-1) \cup (N_0)) \cdot f(z_{N-1}|(N-2) \cup (N_0)) \cdot \\ & \quad \cdots f(z_2|z_1 \cup (N_0)) \cdot f(z_1|(N_0)) \end{aligned} \tag{1}$$

where, e.g., $f(z_{N-1}|(N-2) \cup (N_0))$ represents the conditional pdf of z_{N-1} given the values $\{z_1, z_2, \ldots, z_{N-2}\}$ (denoted by $(N-2)$) and given the set of N_0 original data values.

Thus the generation of a realization by sequential simulation takes the following steps:

1. Draw a value z_1 from the conditional distribution of Z_1 given the data (N_0)
2. Draw a value z_2 from the conditional distribution of Z_2 given that $Z_1 = z_1$ and the data (N_0)

⋮

N. Draw a value z_N from the conditional distribution of Z_N given that $\{Z_1 = z_1, Z_2 = z_2, \ldots, Z_{N-1} = z_{N-1}\}$ and the data (N_0)

Notice that sequential simulation is conditional by construction eliminating the expensive conditioning step required by other traditional techniques, such as turning bands, to transform an unconditional realization into a conditional one (Journel and Huijbregts, 1978). In case that an unconditional realization is needed one should reduce the set of conditioning data to the null set, in which case the first step of the generation would be replaced by drawing from the marginal distribution of Z_1.

Notice also that there is no distinction on the attribute to which each variable refers to, and that there is no restriction on the spatial location of the random variables yielding an algorithm that can be equally applied to generate one or multiple variables either on a regular or an irregular grid.

It remains the problem of determining the conditional probability distribution function of any single random variable given any set of conditioning random variables. This problem has been solved for some multivariate models of the N random variables.

Conditional distributions

The conditional distribution function of a single random variable given any set of conditioning random variables can be computed for some multivariate models of the N random variables (Journel, 1993).

MultiGaussian model

Let $Y_i = \phi(Z_i)$ be the normal score transform of Z_i. If the multivariate distribution of $\{Y_i, i = 1, \ldots, N\}$ is multiGaussian then all conditional distributions of Y_i given any set of conditioning y_j values is univariate Gaussian with mean and variance given by the solution of a set of normal equations (Anderson, 1984) also known as simple kriging equations.

Non-parametric model

The conditional cumulative probability distribution of a single variable Z_i given (n) data can be written as

$$F(z_i|(n)) = E\{I(z_i)|(n)\} \tag{2}$$

with $I(z_i) = 1$ if $Z_i \leq z_i$, $=0$ if not being the indicator transform of the random variable Z_i for threshold z_i. The conditional expectation of the indicator variable can be estimated by indicator kriging (Journel, 1983).

Other models

There are other models for which solution to the problem of computing the cpdfs has been solved. If an isofactorial model is adopted then disjunctive kriging could be used to estimate the cpdfs (Matheron, 1976). If the non-parametric model is used, alternatives to indicator kriging for the estimation of the cpdfs are co-indicator kriging, probability kriging or extended normal equations (Guardiano and Srivastava, 1993).

Practice

Selecting a model for the multivariate probability distribution of the N random variables so that the conditional distribution can be computed solves the theoretical problem of sequential simulation. However, this is not enough from a practical point of view. A number of problems appear at the implementation stage.

Number of conditioning points

The sequential simulation algorithm previously described calls for the computation of conditional probabilities with an increasing number of conditioning values as the algorithm progresses. For any given multivariate pdf model the computation of the conditional probabilities amounts to solving a system of linear equations the number of which is proportional to the number of conditioning points. Therefore, exact application of the algorithm requires the solution of very large systems of equations. To avoid having to solve too large systems of equations and thus render the algorithm feasible, the computation of the cpdf $f(z_i|(n))$, with (n) representing a possibly very large number of conditioning values, is approximated by the cpdf $f(z_i|(n'))$ with $(n') \subset (n)$, $n' \ll n$. The problem is then to find the subset (n') of the original (n) conditioning random variables that best approximates the sought cpdf. In most cases, the subset of size (n') that provides such best approximation is constituted by the conditioning values which are "closest" to Z_i, "closest" being defined in terms of some correlation function.

A practical solution would be as follows: (i) define the maximum number of values that will be used to compute any cpdf (ii) define a neighborhood around the location of Z_i, (iii) retain only the conditioning values within the neighborhood, (iv) sort the retained values in order of increasing "distance" from Z_i as measured by the variogram or a similar measure of structural distance, (v) retain only the closest values up to the established maximum (sometimes is also advisable to make some prior declustering by limiting the maximum number of values to be retained per octant within the search neighborhood, Isaaks and Srivastava, 1989).

Remarks. 1. The search neighborhood should be as large as the maximum distance for which the variograms are to be reproduced. This requirement is difficult to meet in the case of strong geometric anisotropies and in the case of zonal anisotropies.

2. The search neighborhood could be of any shape but it is recommended that its shape be ellipsoidal with the major axes aligned with the major directions of continuity.

3. When all random variables lie on a regular grid, an efficient search algorithm to find out the conditioning values within the search neighborhood is based on a template containing the coordinates of all possible locations within the neighborhood. The template has been initially sorted according to the structural "distance" of the template points to the point being simulated. Looking up the template for the conditioning values is equivalent to a spiral search starting from the location being simulated.

Simulation sequence

The decomposition of the multivariate pdf into the product of univariate cpdfs given in (1) is independent of the ordering of the N random variables. However, because of the simplifications needed for the computation of the cpdfs, a sequence that visits adjacent nodes in a consecutive manner should be avoided in order to reduce the appearance of random artifacts.

It is recommended to apply the algorithm using a random sequence of the N random variables. Such random sequence can be obtained by assigning an integer index to each node and using a random generator of these indices. A congruential generator of the form

$$p_k = (5 \times p_{k-1} + 1) \bmod 2^n$$

will generate a random sequence of integers $\{p_0, p_1, \dots, p_N\}$ containing once and only once each integer value between 1 and 2^n.

Conditioning data

In most cases the N random variables lie on a regular grid. Also, in most cases, the N_0 initial conditioning data do not lie on such regular grid. There are two possibilities on how to treat the conditioning data. One is to relocate each conditioning datum to its nearest grid node. The other one is to keep the conditioning data at their actual coordinates.

Relocating each conditioning datum to its nearest grid node is the most computationally efficient solution. When all the variables are on a regular grid, templates can be built to speed up the search for the nearest conditioning values and to speed up the setup of the systems of equations that must be solved to determine the conditional distributions.

Leaving the data at their actual coordinates hampers considerably the efficiency gained with the templates: the conditioning data values cannot be accommodated by any template, so they have to be treated separately. The spiral search described above must be replaced by slower search strategies.

There is a tradeoff between speed and accuracy between relocating or not the data. Relocating the data increases the simulation speed but it implies loss of accuracy, particularly if the simulation grid is coarse; it may also imply having to discard some of the data because two or more data are relocated to the same grid node, only one of which could be retained. Leaving the data at their locations slows down the simulation but it is the most accurate.

Reproducing zonal anisotropies

Zonal anisotropy is a degenerate case of geometric anisotropy when one of the nested structures modeling the covariance/variogram has an infinite range in one of their major directions. To reproduce properly this type of anisotropy the search neighborhood must be infinitely large in the direction of the infinite range.

The use of a search neighborhood as large as the entire simulation domain is, for simulations on large grids, impractical. The larger the search neighborhood, the slower the search for conditioning data and the larger the amount of computer resources needed. When the reproduction of very long ranges with respect to the grid spacing is important the alternative is to use a multigrid approach.

A multigrid approach is equivalent to using a structured sequence to visit all N random variables. Two or more nested grids are defined over the simulation domain. The finest of these grids coincides with the grid on which all the random variables lie. The visiting sequence starts by the nodes on the coarsest grid. At this stage, since the spacing between grid nodes is large, the search neighborhood can be large in size and include a small number of grid nodes. When the nodes on the coarsest grid have been simulated, the visiting sequence continues to the next finer grid, at the same time reducing the size of the search neighborhood. The sequence goes on until the simulation of the nodes on the finest grid.

From our experience, to reproduce a zonal anisotropy it is enough to use two grids. The coarse one having a very coarse discretization only on the direction of the minimum variance. To reproduce very long ranges in all directions more than two nested grids are needed, the number of them depending on the range values relative to the grid spacing at the finest grid.

Computer codes

Some public domain computer codes implementing sequential simulation for some of the multivariate models discussed before are discused below. They are available from the authors upon request.

ISIM3D (Gómez-Hernández and Srivastava, 1990) and `sisim` (Deutsch and Journel, 1992) are three-dimensional sequential simulation codes for a single attribute. The cpdfs are computed using indicator kriging. ISIM3D relocates the conditioning data to the nearest grid node and allows for two nested grids. `sisim` has the option of preserving the conditioning data at their original locations but it is limited to one single grid.

GCOSIM3D (Gómez-Hernández and Journel, 1993) and `sgsim` (Deutsch and Journel, 1992) are three-dimensional sequential simulation codes for the generation of realizations from a multiGaussian pdf. `sgsim` is limited to the generation of one single attribute whereas GCOSIM3D allows the generation of several attributes.

Evaluation

The algorithm of sequential simulation for the generation of unconditional realizations of a single attribute drawn from a multiGaussian model is evaluated in this section. Conditioning will not be tested because it is automatic: sequential simulation is conditional by construction.

All analyses are carried out on 100 realizations of a two-dimensional field with 50 by 50 grid nodes, the grid spacing is one unit in each direction. The program GCOSIM3D (Gómez-Hernández and Journel, 1993) is used for the generation of the realizations. The only critical parameter that remains constant for all the tests is the maximum number of points to be retained within the search neighborhood which is set to 8.

Reproduction of the multivariate model

A multiGaussian model with zero mean, unit variance and an exponential isotropic variogram with a practical range of 21 is used. The search neighborhood is circular with a radius of 30. A number of tests which are described below are carried out to determine the goodness of reproduction of the multiGaussian model by the simulation algorithm.

Histogram reproduction. Figure 1 shows the histogram computed from the 100 realizations. The zero-mean unit-variance Gaussian histogram is well reproduced.

Variogram reproduction. Figure 2 shows the input model variogram (solid line) and the average variogram computed through the 100 realizations (dotted line). The reproduction is almost perfect.

Indicator variogram reproduction. Since the analytical expression of the indicator variograms for a multiGaussian distribution is known (Xiao, 1985; Journel, 1989) we can compute the experimental indicator variograms and compared them to the analytical ones. Figure 3 shows the theoretical indicator variograms and the experimental ones for thresholds corresponding to the first and last deciles, upper and lower quartiles, and median thresholds. Again the reproduction is almost perfect.

Block variograms. A change of support was applied to the previous 100 realizations using a block of 5 by 5 units. Within the block the arithmetic average was computed and it was assigned to the center of the block. The result is 100 realizations of a variable which should be multiGaussian distributed with zero mean and variogram γ_V given by (Journel and Huijbregts, 1978)

$$\gamma_V(\mathbf{h}) = \overline{\gamma}(V, V_{\mathbf{h}}) - \overline{\gamma}(V, V)$$

where $\overline{\gamma}(V, V_{\mathbf{h}})$ is the average variogram given by

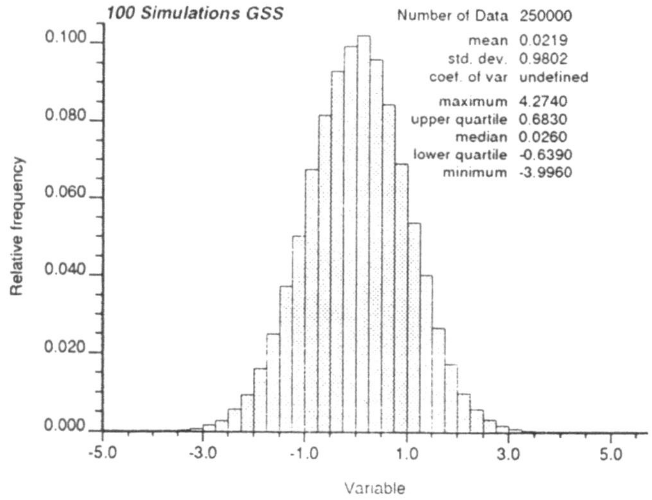

Figure 1: Histogram of the 100 realizations. The model histogram is Gaussian with zero mean and unit variance

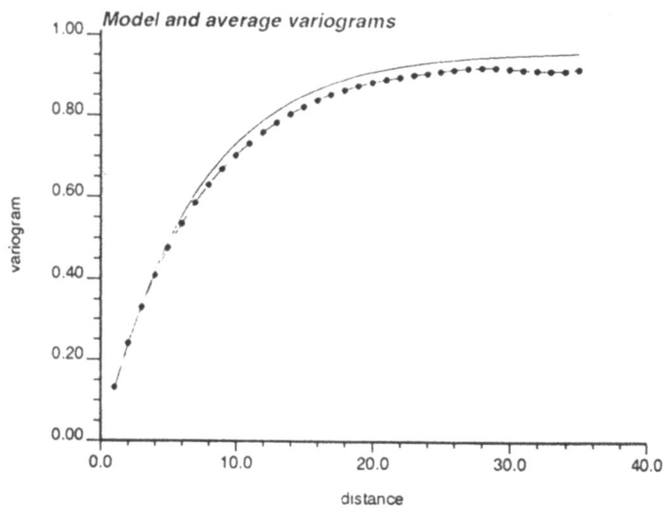

Figure 2: Average variogram versus model variogram. The model variogram (solid line) is exponential with practical range 21. The average variogram is displayed as a dotted line

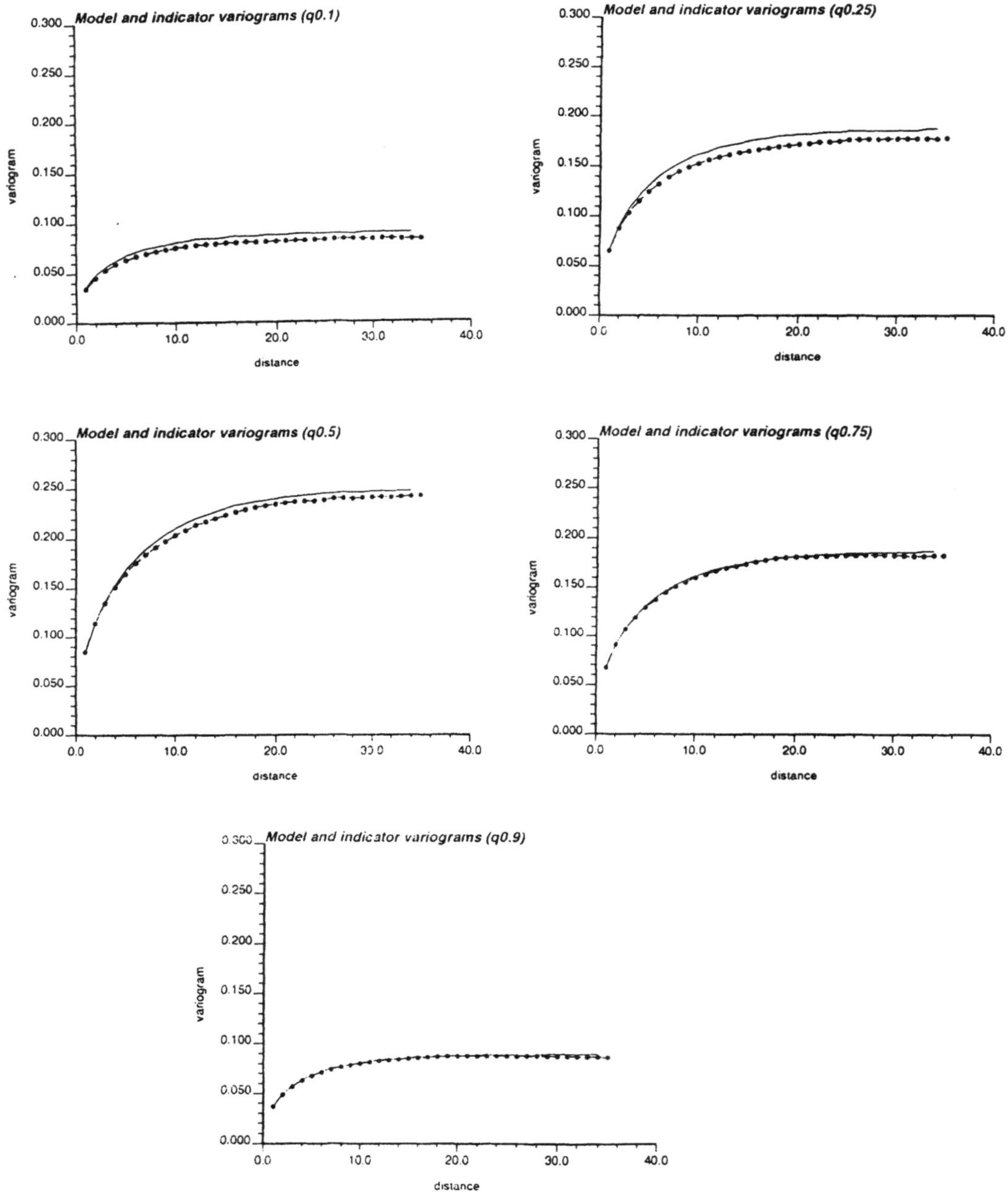

Figure 3: Reproduction of the theoretical indicator variograms. The theoretical indicator variograms (solid lines) for thresholds corresponding to the first and last deciles, lower and upper quartiles and median are compared to the ones obtained from the simulations (dotted lines)

$$\overline{\gamma}(V, V_{\mathbf{h}}) = \int_{\mathbf{x}\in V} \int_{\mathbf{x}'\in V_h} \gamma(\mathbf{x} - \mathbf{x}')d\mathbf{x}d\mathbf{x}'$$

with V representing the block support and $V_{\mathbf{h}}$ representing the same support translated by the vector $\mathbf{h}$. $\overline{\gamma}(V, V)$ is equal to $\overline{\gamma}(V, V_{\mathbf{h}})$ for $\mathbf{h} = 0$. The expression of $\gamma_V(\mathbf{h})$ can be evaluated numerically from the expression of the variogram $\gamma(\mathbf{h})$.
Figure 4 shows the theoretical variogram corresponding to average values taken on blocks 5 by 5 (solid line) and the average variogram computed from the 100 realizations (dotted line). The agreement is very good.

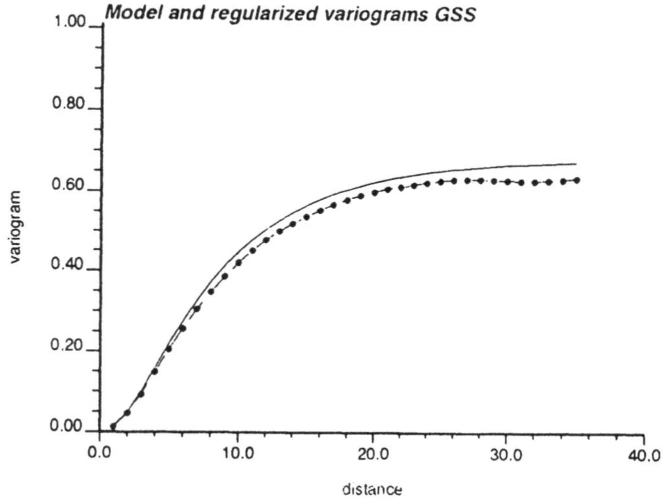

Figure 4: Block variogram. The theoretical regularized variogram for a block of 5 by 5 units correspoding to the same exponential variogram as all previous figures is compared to the one obtained from the same 100 realizations as before after a change of support

The same procedure was applied to blocks of 10 by 10 and the resulting theoretical and experimental variograms are shown in figure 5.
From the result of these tests we conclude that the sequential simulation algorithm produces realizations which follow a multiGaussian model beyond the reproduction of the input mean and the covariance.

Analysis of other variogram models

Three series of 100 realizations for isotropic variograms with a practical range of 12 and a circular search neighborhood with a radius of 24 were generated. For one series the input variogram model was spherical, for another the variogram used was exponential and for the last one, the variogram model was double exponential. Figure 6 shows the reproduction of the input variograms for the three sets of realizations. Both the exponential and the double exponential are well reproduced, whereas the

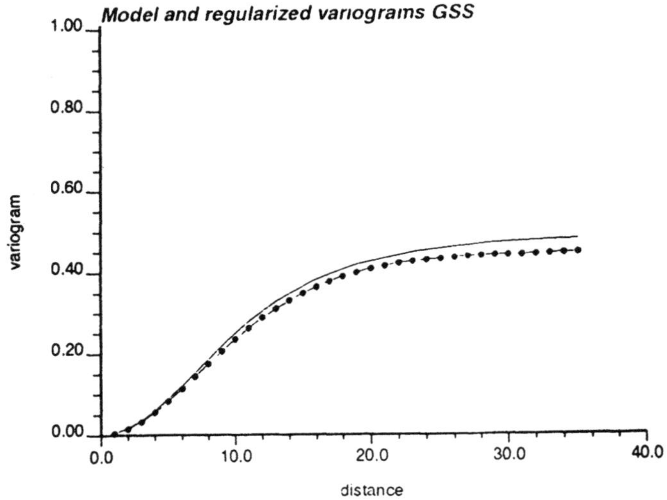

Figure 5: Block variograms. Same as previous figure but using a block support of 10 by 10 units

realizations corresponding to the spherical variogram present a larger range and a shape closer to an exponential variogram.

The Markovian properties of the multiGaussian models with an exponential o a double exponential variogram make the approximation for the cpdf using a small subset of the total number of conditioning values a very good approximation. The multi-Gaussian model with a spherical variogram lacks this property and, as a result, the realizations tend to display an exponential variogram with the same behavior near the origin as the spherical variogram they are trying to reproduce. A better reproduction of the spherical variogram can be obtaining by retaining a larger subset of conditioning values within the neighborhood (only the 8 "closest" points were used for the estimation of the cpdf in this case).

Anisotropic model

An anisotropic exponential variogram model with practical ranges of 21 and 100 in the directions parallel to the cartesian axes is used. The input variogram and the reproduction by the realizations are shown in figure 7. Again, the reproduction is very good.

Sensitivity to the size of the search neighborhood

A final sensitivity analysis is carried out to see the impact of the size of the search neighborhood on the reproduction of an exponential variogram with practical range of 12. Three search neighborhoods are used with radius 12, 24 and 36. Figure 8 shows the input model variogram and the experimental variograms. The reproduction of the variogram is good enough for the smallest search neighborhood.

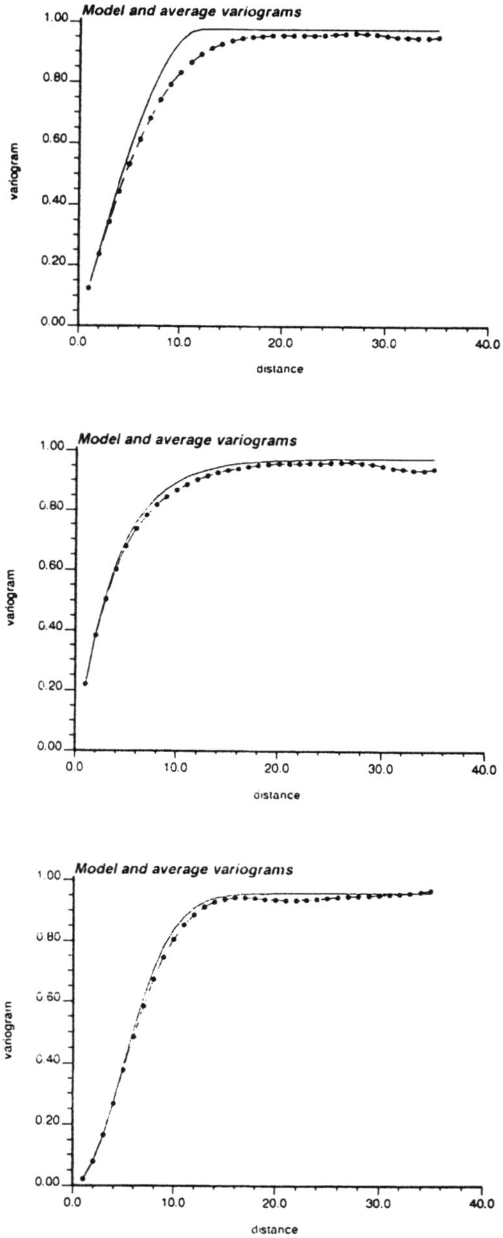

Figure 6: Diferent variogram models. Three different input variogram models with the same practical range are displayed along with the reproduction by a set of 100 realizations. Only the spherical model is not almost perfectly reproduced

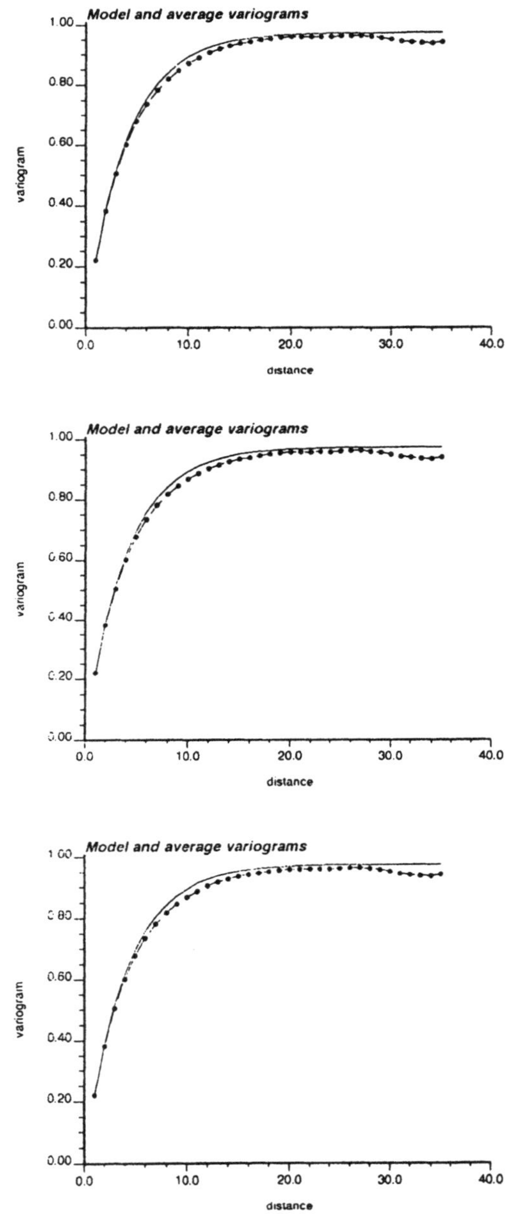

Figure 8: Various sizes for the search neighborhood. Three sizes of the search neighborhood (12, 24 and 36 units) were tested to see their impact on the reproduction of the input variogram. All three sizes yield good reproductions

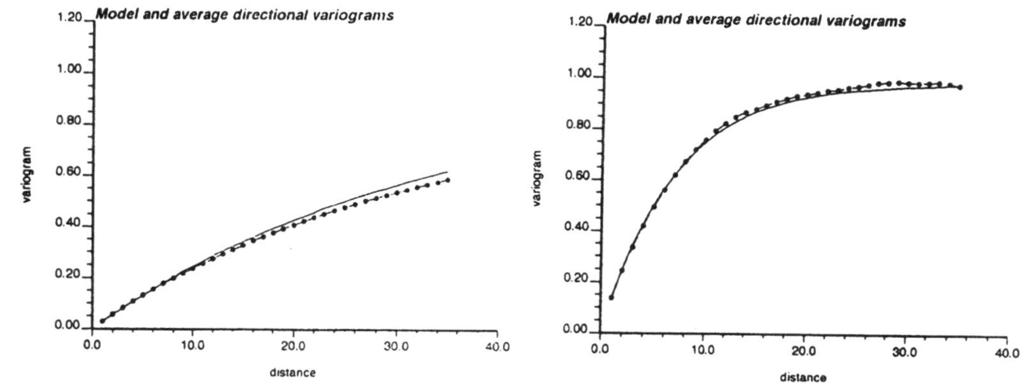

Figure 7: Anisotropic variogram. The reproduction of an anisotropic variogram is also good. The figures show the directional variograms (model and experimental) on the two principal directions

Conclusions

Sequential simulation is a theoretically simple simulation technique which is completely general and conditional by construction. Due to implementation problems some approximations are needed. The impact of these approximations for the case of a multiGaussian model for the random variables to be simulated has been analyzed. The tests show that the approximations do not have a tremendous impact on the reproduction of the underlying multiGaussian model.

Acknowledgments

This research was funded by the Spanish Nuclear Waste Management Company (ENRESA). The opinions expressed on this paper are those of the authors and do not have to coincide with those of ENRESA.

References

Alabert, F. G. (1987). The practice of fast conditional simulations through the lu decomposition of the covariance matrix. *Math. Geology*, 19(5):369–387.

Anderson, T. W. (1984). *Multivariate statistical analysis.* Wiley, New York.

Deutsch, C. V. and Journel, A. G. (1992). *GSLIB, Geostatistical Software Library and User's Guide.* Oxford University Press, New York.

Gómez-Hernández, J. J. and Journel, A. G. (1993). Joint simulation of multiGaussian random variables. In Soares, A., editor, *Geostatistics Tróia '92, volume 1*, pages 85–94. Kluwer.

Gómez-Hernández, J. J. and Srivastava, R. M. (1990). ISIM3D: An ANSI-C three dimensional multiple indicator conditional simulation program. *Computer and Geosciences*, 16(4):395–440.

Guardiano, F. B. and Srivastava, R. M. (1993). Multivariate geostatistics: Beyond bivariate models. In Soares, A., editor, *Geostatistics Tróia '92, volume 1*, pages 133–144. Kluwer.

Isaaks, E. H. and Srivastava, R. M. (1989). *An Introduction to Applied Geostatistics*. Oxford Press, New York.

Journel, A. G. (1983). Non-parametric estimation of spatial distributions. *Math. Geology*, 5(3):445–468.

Journel, A. G. (1989). *Fundamentals of Geostatistics in Five Lessons*, volume 8 of *Short Courses in Geology*. AGU, Washington D.C.

Journel, A. G. (1993). Geostatistics: Roadblocks and challenges. In Soares, A., editor, *Geostatistics Tróia '92, volume 1*, pages 213–224. Kluwer.

Journel, A. G. and Huijbregts, C. J. (1978). *Mining Geostatistics*. Academic Press. London.

Matheron, G. (1976). A simple substitute for conditional expectation: the disjunctive kriging. In *Advanced Geostatistics in the Mining Industry*, pages 221–236, Rome. NATO ASI.

Omre, H., Sølna, K., and Tjelmeland, H. (1993). Simulation of random functions on large lattices. In Soares, A., editor, *Geostatistics Tróia '92, volume 1*, pages 179–199. Kluwer.

Verly, G. W. (1993). Sequential gaussian cosimulation: A simulation method integrating several types of information. In Soares, A., editor, *Geostatistics Tróia '92, volume 1*, pages 543–554. Kluwer.

Xiao, H. (1985). A description of the behavior of indicator variograms for a bivariate normal distribution. Master's thesis, Stanford University.

DISCUSSION FOLLOWING SESSION NO 4

Chairman: Peter Ravenscroft

Papers: Gomez–Hernandez; Haas

The paper by Dr André Haas is not included in these proceedings because a French version of it has already been published under the title "Simulation de réservoirs pétroliers par inversion géostatistique" in the proceedings of the Journées de Géostatistique 25–26 May 93, "Cahiers de Géostatistique, Fascicule 3", Ecole des Mines de Paris, July 93, pp87–100.

As the first presentation had finished on time, the chairman asked whether there were any questions on it.

Allard: Firstly I would like to comment on your first transparency where you show a turning band variogram map. This is not a typical turning band variogram map. It is merely typical of what is obtained using the TUBA software, isn't it?

Gomez: Yes it was obtained using TUBA.

Allard: It is a little unfair to criticize a theoretical algorithm because of faulty software.

Gomez: No I don't think it is faulty. It has been very well accepted, for a long time.

Allard: I know that it is possible, using turning bands, to produce far more acceptable simulations.

Gomez: Of course, you can do much better if you increase the number of lines. The problem is due to the finite number of lines.

Allard: It is not a question of algorithm. It is a question of software. It is not the same.

Gomez: This was generated by TUBA which is a widely accepted software. I did not write it.

Allard: Now concerning the search neighbourhood used, what is the limitation on the number of nodes using a typical computer?

Gomez: It can be very large. In 3D we need 4 entries for each node within the neighbourhood. So you could have 10 000 or 100 000 nodes. The 4 entries are the x, y, z coordinates plus the distance to the centre.

M. Armstrong and P. A. Dowd (eds.), Geostatistical Simulations, 125–129.

Jaquet: Just a comment. I have been using TUBA and it takes 16 turning band lines as a default value.

Gomez: But this was 16 lines. Sixty–four still gives stripes. In this case with 50 x 50 pixel with an integral scale of 7, it took 614 lines in order not to see the striping. So it takes forever.

Jaquet: But it is well-known that you need several hundred lines (and not just 16) in order to simulate properly.

Chilès: You have presented us with a simulation of a multigaussian field, and in this case, provided the variogram has good Markov properties you get a good simulation. Could you comment on the simulation of indicators because in that case you do not have the conditional probability and you have to approximate it by indicator kriging?

Gomez: Indicator simulation is much more complex. More tricks of the trade are required to make it work properly. It does not work as well as the multigaussian which is really well suited to sequential simulation. The first problem is how to estimate the indicator probability, the conditional probability. In principle, you can estimate it conditional to the data using a full cokriging rather than indicator kriging or indicator cokriging. In general this is not done in practice because it requires too many direct and cross indicator variograms. So the conditional probability that the point being simulated is below a given threshold is computed using only the indicator values for that particular threshold. As a consequence, order relation problems appear in the estimated cdf that must be fixed. Then you can draw from the estimated cdf. This drawing only tells you to which class the number belongs. For example you might know that it is between the 3rd and 4th thresholds but you do not have the real number, the real value. The next step is the most problematic. You have to decide how to select a real value within that class.

There are two problems. Firstly for the extreme classes, particularly the upper tail, you have to specify a model for their behaviour. So even though the method is mainly non–parametric, at this point you have to go parametric and say, for example, I want the upper tail of the estimated cdf to look like a Pareto distribution or a linear distribution between the uppermost threshold and some overall maximum. This choice can influence the results dramatically. I think that part of the results shown earlier by Carol Gotway were influenced by this. You never really have enough data to model the upper tail.

Now for all the intermediate classes we generally just draw a number at random in that class. This leads to some loss of correlation, which shows up as a nugget effect when you calculate the Z covariance. Additional fixes are needed to overcome this. For example after doing all the indicator estimation, you can do some gaussian estimation to get a value for z, which respects the covariance. If it falls in the class, you keep it. Otherwise you have to correct it.

The chairman then stopped the discussion to allow Dr Haas to present his paper.

After Dr Haas's talk, the chairman opened the discussion.

Jeulin: Can you weight the data (seismic and well data)?

Haas: Yes, for example, we are going to ask the (seismic) interpreter to produce a confidence map. We think it will be possible to adapt the correlation threshold to this.

Daly: It would like to raise 2 points. First, you quote correlation values of between 0.7 and 0.9 on the wells but I have never got correlations as good as that. You must have very good data.

Haas: You are right. In fact, this type of method cannot be used if you have low correlations. So we need very good data, and very precise preprocessing. Matching seismic and well data is a very crucial step. In general people do not work hard enough on it. If you have time and money, it is possible to improve the matching.

Daly: We are not really interested in simulating acoustic impedance. Porosity is more important. But we know how to study acoustic impedance.
So you need to know that you have a very simplistic region, as well as having very good correlations before the methods are applicable. Every one wants to do these sort of inversions.

Haas: Another point is that later we would like to make a direct simulation of porosity values but this is just the first step. The problem is to model impedances from petrophysical data.

As there were no further questions directly on Haas's paper, the chairman went back to the remaining questions on the previous paper.

Gotway: I would like to comment in response to Chilès' question about sequential indicator simulation. All of its common potential pitfalls were mentioned. Under what circumstances do you think that it is superior to the sequential gaussian method?

Gomez: In my experience, it is very good for categorical variables, that is, if you only want the class variable. You have to be very careful going from the

class variable to the real value.
I prefer to lay out the categories and then in a second pass, to use a gaussian simulation to get the real values in each class rather than drawing at random in that class.

Gotway: Given all the extra trouble caused by chopping the distribution up into classes, what justifies all the extra effort?

Gomez: With indicator simulations you can get characteristics which are not multigaussian. I showed that multigaussian simulations means that you get very small correlations for the extreme thresholds. If you want high correlations at extreme values you cannot use multigaussian techniques. You have to revert to indicators because in that case you can specify the covariances of the last classes independently.

Gotway: What particular characteristics of the realisations reflect the continuity of those extreme values? In our work we considered several data sets, for three of which sequential indicator simulations should have done very well for the reasons that you have just mentioned. The gaussian methods should not have been able to reproduce certain of the visual characteristics of the field, and yet I found that quite the opposite was true.

Gomez: I feel that this was due to the conditioning data that you had. You had quite a lot with interdistances that are small relative to the covariance ranges. You are referring to the Berea data set which has very long ranges for all the covariances. You used 100 data. Whatever is not taken care of by the covariances, will be controlled by the data.

Omre: We have also be done some work on sequential indicator methods using gaussian variograms. Then we get very smooth variogram close to the origin. In our experience this seems to give numerical instabilities. So points which are close, are very highly correlated. When you try to decompose the matrices, it breaks down, not completely giving all 0s or all 1s in your eigenvalues. It seems to give strange numbers. So it is a bad way of breaking down.
So you have to be very careful if you have a double exponential with a long range, because the flat part of the variogram is so long.

Gomez: The point is that the behaviour at the origin is very continuous, even differentiable. The quick fix for that is to add a little nugget effect, say 1%.

Omre: In the correlation matrix, you also have a relation between nearby points.

Gomez: That is, enough to fix it.

Omre: If you have a spherical variogram and a lot of points in your search neighbourhood (say 40, 50, 60), you can have numerical problems. With a very dense grid, the points will all be close to each other.

Gomez: I have not had any problems with the spherical. It is linear at the origin.

Omre: Even for small distances?

Gomez: No, it seems to be different enough.

Haas: But the simple solution is to add a very small nugget effect. Generally that suffices.

Editor: To clarify his comments, Jaime Gomez has added the following remarks:
(1) There will always be at least 1% difference between the values of the off diagonal elements, which is enough to make the instabilities disappear.
(2) As a rule of thumb, for the spherical variogram, as long as the distance between the points is larger than 1/1000th of the correlation range the kriging system is stable. In any case adding a 1% nugget effect fixes any instabilities.

Omre: But when you simulate 100 realisations, you really have to be more than "generally" right. We are not looking at just one simulation. In modern uses, you generate hundreds or thousands of realisations. So we have to be more concerned about these things.

Ravenscroft: In your presentation, you mentioned the possibility of using disjunctive kriging. Have you tried this?

Gomez: No I have not. In principle it is not more complicated; but calculation the conditional probabilities would be more cumbersome.

Ravenscroft: That could address the problems encountered with the indicators.

The chairman then closed the session.

REMINDERS ON THE CONDITIONING KRIGING

Chantal de FOUQUET
Centre de Géostatistique, Ecole des Mines de Paris
35 rue St–Honoré, 77305 FONTAINEBLEAU, France

ABSTRACT

Conditioning simulations by using kriging was proposed by G. Matheron at the start of the 70s. The following presentation only recalls results that should be already well–known.
Simulation methods enable us to construct realizations of RF having a fixed covariance K. In practice, the values z (x_α) at the data points x_α are known and we dictate that at these points the RF Z and its simulations must coincide.
The conditioning method by kriging makes it possible to carry out a conditional simulation into two steps: construction of a nonconditional simulation, then the conditioning kriging. In the general case, simulations obtained in this way have the specified covariance K and are such that at the data points $t(x_\alpha) = z(x_\alpha)$. In addition, for a RF having multigaussian distribution, the conditional distribution $Z(x) \mid Z(x_{\alpha_1}), ..., Z(x_{\alpha_n})$ is reproduced.

1. INTRODUCTION

The reason for generating geostatistical simulations is to provide numerical models that are used as input for empirical calculations such as flow simulations, sensitivity studies, evaluation of a result under constraints. The more the numerical model resembles the actual values, the more reliable are the results. In particular, the numerical model has to restitute the experimental values at the data points: at these points, numerical model and reality coincide exactly. The scope of simulation methods is to extend the spatial structure to the other points to simulate.
Kriging provides the most accurate linear estimation given the data, but the spatial structure of the estimated values differs from that of the actual ones. Thus the kriged values cannot be used as a suitable numerical model when the variability has to be taken into acount (Fig. 1).
In geostatistics the variable y(x) is interpreted as a realization of a RF Y(x). A simulation S is a RF having the same spatial distribution as that of Y, and in particular with

M. Armstrong and P. A. Dowd (eds.), Geostatistical Simulations, 131–145.

- the same expectation and the same variance
- the same covariance or variogram
- the same univariate and bivariate distributions.

The simulation is a conditional one (and is written as T) if, for all data points x_α :

$$T(x_\alpha) = Y(x_\alpha)$$

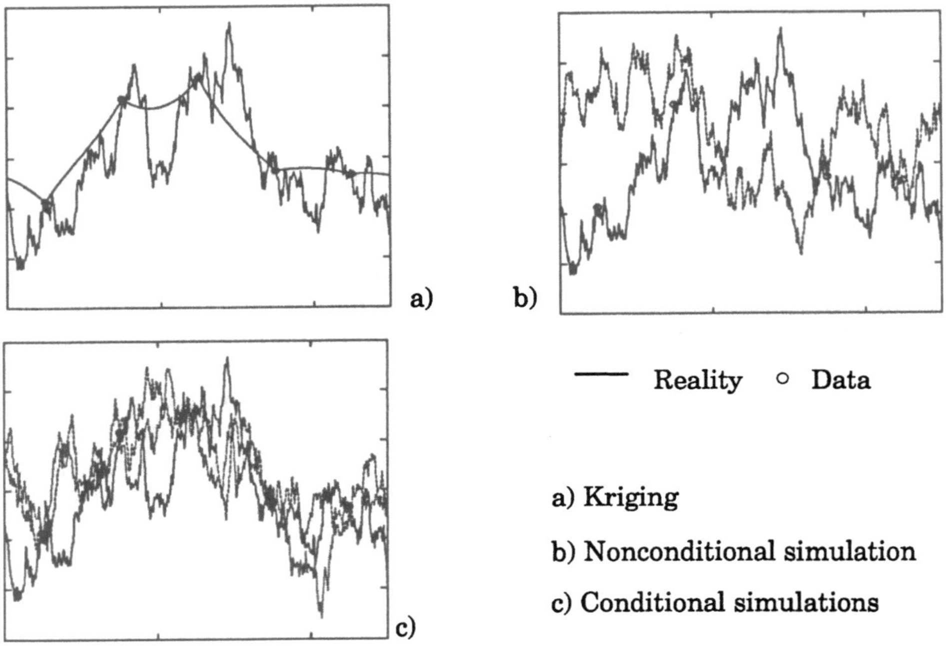

a) Kriging

b) Nonconditional simulation

c) Conditional simulations

Figure 1 Estimation & Simulations

In practice, it is rarely possible to infer more than the bivariate distribution from the data. Then a RF T is often called a conditional simulation of Y, if T has the same bivariate distribution as Y for any pair of points x and x+h.

2. CONDITIONAL DISTRIBUTION : THE GAUSSIAN CASE

Let Z be a RF with spatial distribution F. One realization is known at the experimental points x_α. A conditional simulation is a random function T, having the same spatial distribution as Z, and such that at the experimental points:

$$T(x_\alpha) = Z(x_\alpha)$$

So, at the data points, for any realization of T:

$$t(x_\alpha) = z(x_\alpha)$$

Because of the relation between the joint and the marginal composed probabilities:

$$dF(z_1, ..., z_m \; ; \; z_{\alpha_1}, ..., z_{\alpha_n}) = dF(z_1, ..., z_m | z_{\alpha_1}, ..., z_{\alpha_n}) dF(z_{\alpha_1}, ..., z_{\alpha_n})$$

the T distribution conditioned by the $T(x_\alpha)$ is the conditional distribution

$$dF(z_1, ..., z_m | z_{\alpha_1}, ..., z_{\alpha_n}) = dF(z_1, ..., z_m \; ; \; z_{\alpha_1}, ..., z_{\alpha_n}) / \; dF(z_{\alpha_1}, ..., z_{\alpha_n})$$

In the general case, this conditional distribution is unknown. One well–known special case where this conditional distribution is computable, is that of multigaussian RF. In this case the conditional function noted Z_c is a gaussian RF whose expectation at point x is a linear function of the data $Z(x_\alpha)$ and consequently is equal to the kriging Z^* of Z given the $Z(x_\alpha)$. The residual variance of Z_c is equal to the kriging variance σ_K^2. This gaussian function can be decomposed :

$$Z_c(x) = Z^*(x) + X(x) \qquad (1)$$

where X is a gaussian random function with a zero mean and a nonstationary variance $\sigma_K^2(x)$. The covariance K_ε of X is given by:

$$K_\varepsilon(x, x + h) = K(h) - \lambda^\alpha(x)\lambda^\beta(x + h)K(x_\alpha, x_\beta)$$

where $\lambda^\alpha(x)$ denotes the weight of x_α in the kriging of Z at the point x.
The kriging error X is uncorrelated with Z^*. As the pair (Z, Z^*) is bivariate normal, the pair (Z^*, X) is too, and the uncorrelated RFs Z^* and X are then independent. A conditional simulation T, having the same gaussian spatial distribution as Z, can be decomposed according to (1). The first term is equal to the kriging Z^* given the data $Z(x_\alpha)$ and the second term is a RF R, independent of Z^* and having the same nonstationary distribution as X :

$$T(x) = Z^*(x) + R(x)$$

The direct construction of such a nonstationary RF R is not easy. The conditioning kriging method makes use of the independence of X and R relative to Z^*. R is constructed as a known kriging error, calculated on an nonconditional simulation S independent of Z. This is the "Repiquage des Résidus" method.

3. CONDITIONING KRIGING: THE FAST CASE

Let Y be a stationary RF with a zero expectation and a given covariance K. The set of data points $\{x_\alpha\}$ is assumed finite.

3.a. Construction of conditional simulations

Y(x) is the sum of two RF without mutual correlation: the kriged values $Y^*(x)$ and the error $\varepsilon(x)$. The error is unknown on the realization:

$$Y(x) = Y^*(x) + (Y(x) - Y^*(x)) = Y^*(x) + \varepsilon(x) \qquad (2)$$

Let S be an independent simulation of Y. The realizations of S are constructed at the points x_α and at other x points. As for Y, we can krige S at any point x **using only the values at points x_α.** In the decomposition:

$$S(x) = S^*(x) + (S(x) - S^*(x)) = S^*(x) + \varepsilon_s(x) \qquad (3)$$

the variable S and its kriging S^* are known for all realizations; so on the simulation S, the kriging error is known.
Let T be the RF constructed, replacing in the formula (2), the error ε that is independent of Y^* by the error ε_s calculated on a nonconditional simulation.

$$T(x) = Y^*(x) + (S(x) - S^*(x)) \qquad (4)$$

As S is independent of Y, the error ε_s is independent of Y^*.

- **Y and T have the same stationary covariance K.**

As the kriging of Y and S are carried out with the same covariance K and with the same configuration of the data points x_α and of the point to be kriged x, the kriging weights $\lambda(x)$ are identical. Consequently the nonstationary covariance of the kriged values $K^*(x\ ,\ x+h)$ is identical for Y^* and S^*. As, using the absence of correlation between the kriging and the error,

$$K(h) = K^*(x\ ,\ x+h) + K_\varepsilon(x, x+h) \qquad (5)$$

the covariance of the errors is equal for $Y - Y^*$ and for $S - S^*$. Then, from the independance of Y and S:

$$E[T(x)T(x+h)] = K^*(x, x+h) + K_\varepsilon(x, x+h) = K(h)$$

- **Y and T have the same value at the data points.**

At the experimental points, the kriging error $\varepsilon_s(x_\alpha)$ is zero on the nonconditional simulation:

$$S(x_\alpha) - S^*(x_\alpha) = 0$$

and for the RF Y :

$$Y(x_\alpha) = Y^*(x_\alpha)$$

Consequently:

$$T(x_\alpha) = Y(x_\alpha)$$

- **Gaussian case: Y and T have the same spatial distribution**

By construction, T is the sum of two independent RFs, respectively with covariances K^* et K_ε.
In the gaussian case, the kriging Y^* and the kriging error $Y-Y^*$ are two independent RFs, each with a multigaussian distribution and with respective

nonstationary covariances K^* et K_ε. If the nonconditional simulation S has a gaussian spatial distribution, then the simulation T, being the sum of two nonstationary gaussian RF, has a spatial gaussian distribution identical to that of Y. T is then a conditional simulation of Y.
In the general case, the distribution of T constructed using (5) would not be identical to that of Y.

3.b. Practical construction

In practice, only one kriging is carried out:

$$T(x) = S(x) + \sum_\alpha \lambda^\alpha(x)[Y(x_\alpha) - S(x_\alpha)]$$

To sum up, the conditional simulation method is the following (Figure 2):

- nonconditional simulation at points x_i **and at the data points** x_α.
- at these points x_α, computation of the difference $y_\alpha - s_\alpha$
- kriging at points x_i, with the covariance K, of the difference $y^\alpha(x_i) - s^*(x_i)$.
- summation of the nonconditional simulation $s(x_i)$ and of the kriging of the difference $(y^* - s^*)(x_i)$.

This procedure enables us to construct any number of conditional simulations.

Remarks :
– As Y is a RF with a zero expectation, this kriging is carried out **with no** unbiasedness conditions.
– The hypothesis of stationarity is unnecessary. Conditioning kriging works if the covariance K(x, x+h) depends on points x et x+h separately, provide that an unconditional simulation method accepting this covariance available of a RF, and in practice that the inference of the non stationary covariance is possible.
– The support of the simulated variable is not necessarily identical to that of the data. When these supports are different, the nonconditional simulation should be carried out in a consistent way at the x_i points for the support chosen for the simulation, and at the x_α points for the support corresponding to that of the data; this requires having a consistent bivariate model at one's disposal. This bivariate model is used in the conditioning kriging system.
Conditioning with change of support will be carried out in the frame work of the discretized Gaussian model.
– As the multigaussian distribution is compatible with any covariance, the conditioning process is quite general under this hypothesis.

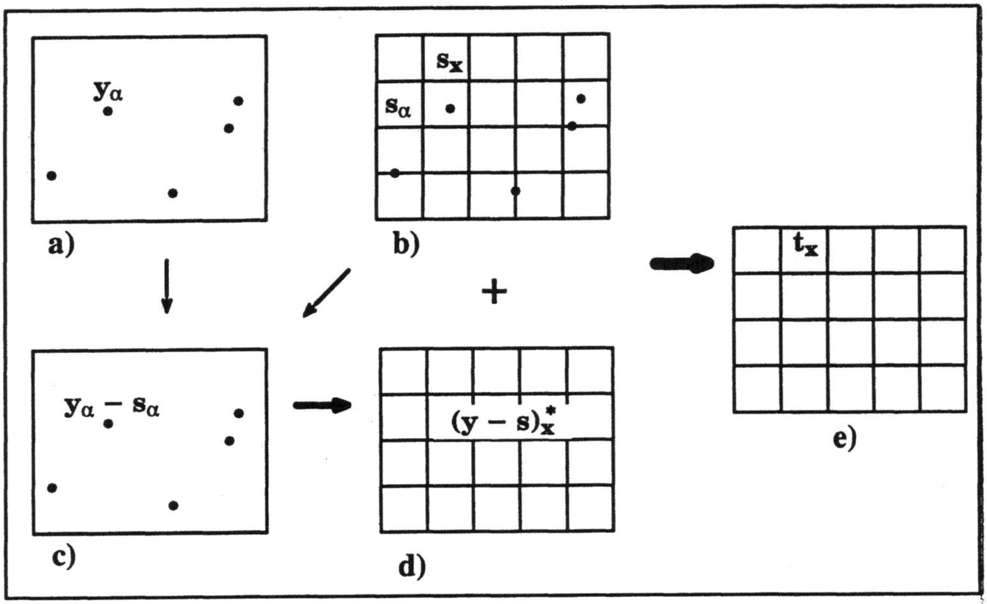

a) data points y_α

b) nonconditional simulation on "grid" and on data points : s_x and s_α

c) calculation of the difference on data points $y_\alpha - s_\alpha$

d) kriging of the difference on "grid" $(y - s)_x{}^*$

e) summing s_x and $(y - s)_x{}^*$

Figure 2 Steps in the construction of a conditional simulation

3.c. Properties of conditional simulations (Fig. 3)
Let Y_α denote the vectorial variable Y_{α_1} , ... , Y_{α_n} and T a conditional simulation constructed from formula (5). At a data point $T(x_\alpha) = Y(x_\alpha)$. Thus the krigings of Y or T using only these data points verify

$$\forall x \quad , \quad T^*(x) = Y^*(x)$$

Expectation
From the decomposition (5) and using the independence of Y and S

$$E(T(x)|Y_\alpha) = E(Y^*(x)|Y_\alpha) + E(\varepsilon_s(x)|Y_\alpha) = Y^*(x)$$

So for T, Y^* summarizes thus all the information brought by the Y_α.. Conditionally to the Y_α, the expectation of a conditional simulation at any point is equal to kriging at this point. Consequently at any point the mean of a large number of realizations is equal to kriging.

As Y^* is the conditional expectation of T when the Y_α are known

$$E(T(x)|Y^*) = Y^*(x)$$

• Gaussian case : Equality of the conditional expectations of Y and T
When Y is gaussian the conditional expectation is equal to kriging, and then

$$E(T(x)|Y_\alpha) = E(Y(x)|Y_\alpha) \qquad (6)$$

The conditional expectations in relation to the Y_α are therefore identical for Y and for any conditional simulation T and are equal to $Y^*(x)$. Consequently

$$E(T(x)|Y^*) = E(Y(x)|Y^*)$$

These properties are a consequence of the identity of the spatial distribution of Y and T.
• When the distribution of Y is not gaussian, generally

$$E(Y(x)|Y^*(x)) \neq Y^*(x)$$

and consequently

$$E(T(x)|Y_\alpha) \neq E(Y(x)|Y_\alpha) \text{ and also } E(T(x)|Y^*) \neq E(Y(x)|Y^*)$$

Remark : The a priori expectation of T is equal to that of Y

$$E[T(x)] = E[E(T(x)|Y^*)] = E(Y^*(x)) = E(Y(x))$$

Variance of conditional simulation
From the decomposition (5),

$$Var[T(x)|Y^*] = Var[T(x)|Y_\alpha] = E[T(x) - Y^*(x)]^2 = \sigma_K^2(x) \qquad (7)$$

Conditionally to the Y_α the variance of T is equal to the kriging variance. The more numerous and better distributed are the data, the better conditioned is the simulation. Otherwise weak conditioning permits wide variations from one realization to another.
The preceding formula also shows that the variance of a large number of realizations at a point is, in expectation, equal to the kriging variance.
In this way, generating numerous conditional simulations provides possible images of the variable being studied but does not give any additional information compared to kriging.

Remark :
Let I_x be the interval $\left[Y_x^* - 2\sigma_k(x), Y^*(x) + 2\sigma_K(x)\right]$. At any point, as T(x) is a gaussian variable with expectation $Y^*(x)$ and variance $\sigma_k^2(x)$, the probability $P(T(x) \in I_x)$ is equal to 0.95. The confidence interval is valid point by point, but does not remain so in terms of vectors or Random Functions. Indeed if I_1 et I_2 are the intervals associated with points x_1 et x_2,

$$P(T(x_1) \in I_1 \ , \ T(x_2) \in I_2) = P(U_1 \in [-2 \ , \ +2], U_2 \in [-2 \ , \ +2])$$

where U_1 et U_2 are two gaussian variables whose covariance is that of the variables:

$$\frac{Y(x_1) - Y_1^*}{\sigma_1} \quad \text{and} \quad \frac{Y(x_2) - Y_2^*}{\sigma_2}$$

Accuracy of a conditional simulation
Let T be a conditional simulation of Y. Let us take T as an estimator of Y. As was shown in (7) the estimation is unbiased:

$$E(Y(x) - T(x)) = 0$$

Because of the decomposition (5) and the independence of Y and Y_s, the estimation variance of Y by T is:

$$D^2(Y(x) - T(x)) = \mathrm{Var}(Y(x) - Y^*(x)) + \mathrm{Var}(Y_s(x) - Y_s^*(x)) = 2\sigma_K^2(x) \qquad (8)$$

The estimation variance by a conditional simulation is double the kriging variance: a simulation does not set out to provide a precise estimate.
The intensity of the conditioning is given from the map of kriging standard deviations. If simulations have to be close to reality everywhere, the domain to be simulated must be restricted to the well informed areas.

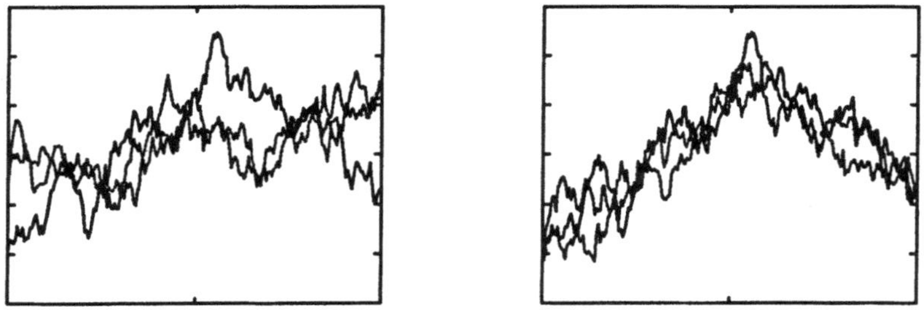

a) Weak conditioning b) Double the conditioning points
—— Reality o Data

Figure 3 Intensity of conditioning

Remarks :
– It can be shown that near the data points, the covariance is a "mixture" of the covariance of the conditioning data and the model K. When many data are available the conditioning "distorts" the structure of the nonconditional simulation and displays that of the data locally. In the multivariate case, strict conditioning can permit to reproduce approximatively complex relationships between the variables on the conditional simulations, thus avoiding difficult modeling of the cross covariances.
– The covariance of Y and T is given by:

$$E[Y(x)T(x + h)] = E(Y^*(x)Y^*(x + h))$$

and

$$E[Y(x)Y * (x + h)] = E[T(x)Y * (x + h)] = E[Y * (x)Y * (x + h)]$$

4. THE HYPOTHESIS OF MULTIGAUSSIAN DISTRIBUTION

Using kriging to do the conditioning is conditioning in the spatial distribution sense only if

$$E(Y|Y_\alpha) = Y^*$$

and if the RF $Y - Y^*$ is **independent** of Y^*. In the other cases the distribution of T differs from that of Y : for an arbitrary RF, the univariate and bivariate distributions would not be reproduced by this procedure.
For example, for the indicator of a random set A:

$$Y(x) = \begin{cases} 1 & \text{if } x \in A \\ 0 & \text{otherwise} \end{cases}$$

the kriging error is either $-Y^*(x)$, or $1-Y^*(x)$. It is clear that the error calculated on an nonconditional simulation S that is independent of Y, and that is either $- S * (x)$ or $1 - S * (x)$, will not give the value 1 or 0 at any point by the decomposition (5).
It is not enough for Y(x) to be gaussian, point by point. For example let Y_1 have a multigaussian distribution with an exponential covariance and Y_2 a mosaic distribution, with a gaussian histogram and an exponential covariance. Figure 4 shows that the conditioning procedure gives an image analogous to the initial model in the first case, but not in the second one.

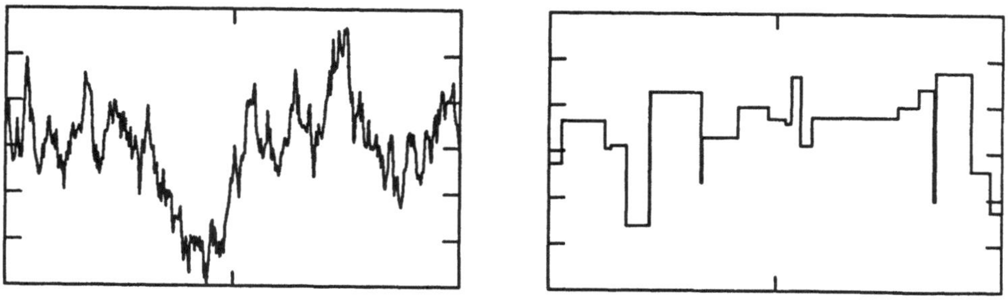

a) Multigaussian RF b) Mosaic RF

Figure 4: 2 RF having same histogram & covariance

Gaussian anamorphosis
In practice, the histogram of the samples z_α is not gaussian. The RF Z is supposed to have a univariate distribution function F. For an arbitrary distribution F, the anamorphosis ϕ, determined by:

$$z = \phi(y) \text{ iff } F(z) = G(y)$$

(where G denotes the cumulative distribution function of the standard normal distribution) transforms a gaussian R.F. into a R.F. with the distribution F.
In order to reproduce the histogram, the conditional simulation T is performed on gaussian values and is transformed into a RF with distribution F using the anamorphosis ϕ by putting

$$T_Z = \phi(T)$$

The anamorphosis ϕ can be expanded in terms of normed Hermite polynomials $H_n(Y)$:

$$Z = \phi(Y) = \sum_0^{\infty} \psi_n H_n(Y)$$

If T is bi–gaussian with covariance K, the covariance C of T_Z is then :

$$C(h) = \sum_1^{\infty} \psi_n^2 K^n(h) \qquad (9)$$

In order to obtain a simulation T_Z with the covariance C, the covariance used for the gaussian simulation (nonconditional simulation S and conditioning kriging) is the covariance K deduced from C by the above relation. The conditioning values for the gaussian conditional simulation are the Y_α defined by

$$Z_\alpha = \phi(Y_\alpha)$$

If the distribution F is a continuous one, then ϕ is bijective, and the Y_α are given by $Y_\alpha = \phi^{-1}(Z_\alpha) = G^{-1}[F(Z_\alpha)]$. The article by X. Freulon in this volume shows how to obtain conditioning data Y_α when ϕ is not bijective. In the following, ϕ is assumed bijective.

To summarize, the steps for constructing a conditional simulation T_Z of Z are the following :

- determining the conditioning values, Y_α, for the gaussian simulation
- calculating the experimental variogram of the Y_α and fitting a model $K(0) - K(h)$
- conditional simulation T of Y with covariance K
- transforming into a conditional simulation of Z : $T_Z = \phi(T)$

The spatial distribution of T_Z is an anamorphosed multigaussian distribution. As previously, only the bivariate distribution of Z is verified. Instead of checking directly the consistency with the anamorphosed bigaussian distribution, it is much easier to check the binormality of the gaussian transformed Y_α.

Given the data Z_α the conditional distribution of T_Z is no longer equal to kriging :

$$E(T_Z \mid Z_\alpha) = E(\phi(T) \mid Y_\alpha) = E[\phi(Y^* + \sigma_k N)]$$

where N is a nonstationary gaussian RF, independent of Y^*. Generally we also have

$$E(T_Z \mid Z_\alpha) \neq E(T_Z \mid Z^*)$$

Remark :

– In practice ϕ is determined from the data z_α. Without a relation defining the conditions of compatibility of a covariance with a given marginal distribution, K is determined first by fitting the experimental variogram of the y_α, the gaussian transforms of z_α. The covariance C deduced from K using formula (9) is then inevitably compatible with the distribution F of Z.

– In practice, the inversion of ϕ can be avoided using, for example, a simple classification of the data z_α. If N data points are available and ranked in increasing order $z_1, \ldots, z_N$, the gaussian value attributed to z_n is simply $y_n = G^{-1}\left(\frac{n}{N+1}\right)$ and we check that $z_n = \phi(y_n)$

Checking the hypothesis of binormality on the conditioning data
As was previously mentioned, it is easier to check the binormality of the y_α than the compatibility of the z_α with the anamorphosed bigaussian model.
Assuming binormality, the following properties are verified :

- the differed scatter diagram Y_x, Y_{x+h} should be elliptical. The regression of one variable on another is linear. The equations for regression lines are:

$$E(Y_x \mid Y_{x+h}) = \varrho Y_{x+h} \text{ et } E(Y_{x+h} \mid Y_x) = \varrho Y_x$$

where ϱ denotes the correlation coefficient between Y_x and Y_{x+h} :

$$\varrho = K(h)$$

- let $\gamma_y(h)$ be the model fitted to the experimental variogram of the y_α et K the associated covariance: $\gamma_Y = K(0) - K(h)$. The experimental variogram of the z_α is fitted by the variogram $\gamma_z(h)=C(o)-C(h)$ where C is deduced from K in accordance with formula (9).
- the ratio of the square root of the variogram of order 2 to the variogram of order 1 (expectation of the absolute value of the deviations : $\gamma_1(h) = \frac{1}{2}E(|Y_{x+h} - Y_x|)$) is constant and equal to $\sqrt{\pi}$:

$$\frac{\sqrt{\gamma(h)}}{\gamma_1(h)} = \sqrt{\pi}\ , \quad h \neq 0$$

In the practical applications, we check whether the gaussian y_α deduced from the z_α verifies these properties. If the bigaussian assumption is apparently unacceptable, we look for other models which are better suited (study of the edge effect, etc.).

5. COMPARISON WITH SEQUENTIAL METHODS

Relation 2 gives another way to construct simulations, conditional or otherwise. In this method, the simulation is constructed point by point, that is in a sequential way.
Let $x_{\alpha_1}, \dots, x_{\alpha_n}$ be the data points. At these points the simulation T is taken as $T(x_\alpha) = Y(x_\alpha)$. Suppose the simulation T has already been constructed at the points $x_1, \dots, x_n$. To construct T at x_{n+1} we simply put

$$T(x_{n+1}) = T^*(x_{n+1}) + \sigma_{n+1}\varepsilon_{n+1}$$

where $T^*(x_{n+1})$ denotes the kriging at x_{n+1} using the data points and the points already simulated, σ_{n+1} is the standard deviation, and ε_{n+1} is a random variable with expectation 0 and unit variance, independent from the other already known random variables.
The properties of this simulation method are analogous to those of the conditioning kriging. For a St RF Y with an arbitrary distribution only the moments of order 1 and 2 are restituted. For example if the Y_α and the ε_i have a uniform distribution, $T(x_n)$ is the sum of uniform variables whose number increases at each step, and then T is not stationary. If the Y_α are multigaussian, and if the ε_i are gaussien, T is a multigaussian RF.
In this method the kriging neighborhood increases by one at each step. Apart from some particular cases with screen effect, such as the exponential covariance in 1D, the kriging configuration using the data points and the already simulated points is modified at each step in the sequential method. In two dimensions, in an unique neighborhood, the first method uses only one inversion of the matrix instead of n for the sequential one, where n is the number of the points to be simulated. In a moving neighborhood, the neighborhood remains the same for groups of points to be simulated in the first method, whereas it differs for each point in the sequential one.
The sequential method is interesting in a few particular cases, such as the autoregressif models in 1D, or for a factorized covariance in 2D.

6. CONDITIONING KRIGING IN IRF−K

Reminder: Kriging equations in IRF−k

The monomials of degree l are denoted by f^l and the admissible linear combinations of the IRF−k by $Z(\lambda)$. To simplify, Z is assumed to have no drift: $E(Z(x)) = m$. Let K be a generalized covariance of Z. It is determined up to an even polynomial with degree lower than or equal to 2k. We know that for admissible linear combinations of Z, all these generalized covariances are equivalent.

The kriging system of $Z(x_0)$ by the Z_α is expressed as:

$$\begin{cases} \lambda^\beta K_{\alpha,\beta} - \mu_l f_\alpha^l = K(x_\alpha - x_0) \\ \lambda^\alpha f_\alpha^l = f_0^l \end{cases} \tag{10}$$

The kriging error is an admissible linear combination with variance:

$$\sigma_k^2(x) = K(0) + \mu_l f^l(x_0) - \lambda^\alpha K(x_\alpha - x_0)$$

In particular when k=0, the system reduces to the following

$$\begin{cases} \lambda^\beta K_{\alpha,\beta} - \mu = K(x_\alpha - x_0) \\ \lambda^\alpha = 1 \end{cases}$$

with

$$\sigma_k^2(x) = K(0) + \mu - \lambda^\alpha K(x_\alpha - x_0)$$

Kriging is an exact interpolator: at a data point, kriging is equal to the value at this point:

$$Z_\alpha^* = Z_\alpha$$

The kriging error is uncorrelated with any admissible linear combination of the data. Indeed let $Z(v)$ be this linear combination of the Z_α ; as seen in (10)

$$E[Z(v)\ (Z_0^* - Z_0)] = v^\alpha \lambda^\beta K_{\alpha,\beta} - v^\alpha K(x_\alpha - x_0)$$

$$= v^\alpha[\mu_l f_\alpha^l + K(x_\alpha - x_0) - K(x_\alpha - x_0)]$$

$$= \mu_l[v^\alpha f_\alpha^l] = 0$$

Thus any admissible linear combination $Z(v)$ can be decomposed into two uncorrelated terms:

$$Z(v) = Z(v)^* + [Z(v) - Z(v)^*]$$

If $Z(v)$ is an admissible linear combination, the same weighting function applied to the Z^* is still an ALC, with the notation $Z^*(v)$. Indeed, let us assume $v^\beta f_\beta^l = 0$. A linear combination $v^\beta Z^*_\beta$ is a linear combination $v^\beta \lambda_\beta^\alpha Z_\alpha$ of the Z_α whose coefficients satisfy $v^\beta \lambda_\beta^\alpha f_\alpha^l = v^\beta f_\beta^l = 0$. Because of the linearity of kriging, $Z^*(v) = [Z(v)]^*$. To simplify the notation, put $\varepsilon(v) = Z(v) - Z(v)^*$.

The generalized covariance K can be decomposed into 2 nonstationary generalized covariances using a formula analogous to (4). Indeed:

$$E[Z(v_x)Z(v_{x+h})] = E[\ (Z^*(v_x) + \varepsilon(v_x))\ (Z^*(v_{x+h}) + \varepsilon(v_{x+h}))\ ]$$

$$= E[Z^*(v_x)Z^*(v_{x+h})] + E[\varepsilon(v_x)\varepsilon(v_{x+h})]$$

If λ_x^α denotes the weight allocated to the location x_α. The nonstationary covariance of the ALC of the Z* is:

$$K^*(v_x, v_{x+h}) = E[Z^*(v_x)Z^*(v_{x+h})] = E[(\sum_\beta v^\beta \sum_\alpha \lambda_{x_\beta}^\alpha Z_\alpha)(\sum_{\beta'} v^{\beta'} \sum_{\alpha'} \lambda_{x_{\beta'}+h}^{\alpha'} Z_{\alpha'})]$$

$$= \sum_{\beta,\beta'} v^\beta v^{\beta'} \sum_{\alpha,\alpha'} \lambda_{x_\beta}^\alpha \lambda_{x_{\beta'}+h}^{\alpha'} K(x_\alpha - x_{\alpha'}) \qquad (11)$$

K is the sum of 2 nonstationary covariances, the covariance of the ALC of the Z* and the covariance of the kriging errors, which is written in its abridged form:

$$K_v(h) = K^*(v_x, v_{x+h}) + K_\varepsilon(v_x, v_{x+h}) \qquad (12)$$

Conditioning kriging & properties

Let Z be an IRF−k, having K as one generalized covariance. Z is known at the experimental points x_α. Let S be an nonconditional simulation of Z, i.e. a IRF of same order k as Z and in the same family of generalized covariances. S is assumed independent of Z and is known at the experimental points x_α and at any of the points x. A conditional simulation T of Z is obtained from S, following exactly the same construction as in the stationary case by putting :

$$T(x) = Z^*(x) + [S(x) - S^*(x)]$$

It is easy to verify that at the experimental points $T(x_\alpha) = Z(x_\alpha)$

As the krigings of Z and S are performed using the same configuration of the experimental points x_α and with equivalent generalized covariances, the weights are identical, and from (11) the nonstationary covariances of ACL of kriged values are also identical. From (12), the nonstationary covariances of the respective errors are identical for Z and any nonconditional simulation S. The indeterminacy on the covariance of S is filtered since we restrict ourselves to using only admissible linear combinations.

As Z and S are assumed independent, the covariance of T is the sum of the two nonstationary covariances of $Z^*(x)$ and of $\varepsilon_S(x)$:

$$K_T(v_x, v_{x+h}) = K^*_Z(v_x, v_{x+h}) + K_{\varepsilon_S}(v_x, v_{x+h}) = K^*_Z(v_x, v_{x+h}) + K_{\varepsilon_Z}(v_x, v_{x+h}) = K_v(h)$$

For any ALC v, T has the same generalized covariance as Z. Consequently T is a IRF−k, having the generalized covariance K.

As T has the same generalized covariance as Z and passes through the values of Z at the experimental points, it is called a **conditional simulation** of Z. The conditioning here is reduced to the equality of the first two moments and to the identity of the RF at the experimental points.

Remark: The conditioning points should be sufficiently numerous for the kriging to have a solution. This ensures that Z and T correspond to the same representation of the IRF−k even if the nonconditional simulations S can be any representation of the IRF−k.

Properties of conditional simulations:
They are analogous to those (6,7,8) obtained in the stationary case:
– Expectation of T conditionally to the $Z(x_\alpha)$: $E(T(x)|Z_\alpha) = Z^*(x)$
As by construction

$$E(T(x)|Z^*) = Z^*(x)$$

generally

$$E(T(x)|Z_\alpha) \neq E(Z(x)|Z_\alpha) \text{ and likewise } E(T(x)|Z^*) \neq E(Z(x)|Z^*)$$

– Variance of T conditionally to the $Z(x_\alpha)$: $Var(T(x)|Z^*) = \sigma_k^2(x)$
– Precision of a conditional simulation T, considered as an estimator of Z

$$Var(Z(x) - T(x)) = 2\sigma_K^2(x)$$

As in the stationary case, the mean of a large number of conditional simulations gives the kriging, and the variance of these simulations around the kriging is equal to the kriging variance.

REFERENCES

JOURNEL A.G., 1974. Simulations conditionelles, théorie & pratique. Thèse de Docteur−Ingénieur, Nancy I

CHILES J.P., 1977. Géostatistique des phénomènes non stationnaires (dans le plan).Thèse de Docteur−Ingénieur, Nancy I

MATHERON G., 1978. L'estimation globale des réserves récupérables. Note C 75. ENSMP, Fontainebleau

MATHERON G., 1978. Le krigeage disjonctif et le paramètrage des réserves Note C 76. ENSMP, Fontainebleau

MATHERON G., 1988. The internal consistency of models in geostatistics. Proceedings of the 3rd International Geostatistical Congress, Sept 3−9 1988, Avignon (France). M. Armstrong, Ed. Kluwer Acad. Press.

NON CONDITIONAL SIMULATION OF STATIONARY ISOTROPIC MULTIGAUSSIAN RANDOM FUNCTIONS

Christian LANTUÉJOUL
Centre de Géostatistique
Ecole des Mines
35 rue Saint-Honoré
77305 Fontainebleau
France

ABSTRACT. In this paper, the problem of simulating a 3-dimensional stationary, isotropic multigaussian random function with covariance C is considered. On the basis of the Central Limit Theorem, a possible procedure consists of simulating a large number of independent stationary random functions (not necessarily multigaussian) with covariance C. This procedure raises two questions:
i) how to simulate a random function of covariance C? Several algorithms are possible (spectral, dilution, turning bands, tessellation, migration ...). These algorithms are briefly presented and compared from various standpoints such as their range of validity and their efficiency.
ii) what is the number of random functions that have to be simulated? Even if no magic number can be recommended, a helpful tool to answer this question is the Berry-Esséen theorem.

1 Introduction

A random function is said to be multigaussian if any linear combination of its variables follows a gaussian distribution. In the stationary case, a multigaussian random function (MGRF) has its spatial distribution totally characterized by its mean value m and its covariance C. To simplify notations, we shall assume $m = 0$ and $C(0) = 1$ (standard MGRF).
This paper is devoted to the non-conditional simulation of stationary, isotropic multigaussian random functions over a discrete or continuous three-dimensional domain.

In the case where the simulation domain is discrete, a sequential procedure can be considered (Alabert, 1987). This consists of prescribing an arbitrary ordering of all of the points of the domain, and of simulating each point in turn according to a conditional gaussian distribution given the generated values of all the previous points.

M. Armstrong and P. A. Dowd (eds.), Geostatistical Simulations, 147–177.

Though this procedure is theoretically sound, it suffers in practice from at least two impediments. Firstly, if the simulation domain has too many points (say more than 500), the procedure cannot be applied directly. Instead, the simulation of each point requires a selection from among the already generated values. The definition of a selection criterion is complicated as it involves the geometry of the already simulated points as well as the covariance type. Such selections result in an error that will propagate in subsequent simulations. Its importance has to be assessed. Secondly, the procedure is numerically unstable if the covariance has a smooth behaviour at the origin.

In the case where the simulation domain is continuous, a 'parallel' procedure is necessary. Let $Y_1, ..., Y_n, ...$ stand for a sequence of standard independent and identically distributed random functions with covariance C. The spatial distribution of the random function

$$Y^{(n)} = \frac{Y_1 + \cdots + Y_n}{\sqrt{n}}$$

tends to become multigaussian with covariance C as n becomes very large. In fact, any linear combination of variables of $Y^{(n)}$ can be written as a linear combination of n independent and identically distributed variables

$$\sum_{j=1}^{p} \lambda_j \, Y^{(n)}(x_j) = \frac{\sum_{j=1}^{p} \lambda_j \, Y_1(x_j) + \cdots + \sum_{j=1}^{p} \lambda_j \, Y_n(x_j)}{\sqrt{n}},$$

which implies, according to the Central Limit Theorem (Feller, 1971), that its distribution tends to become gaussian as n becomes very large.

Two types of questions are encountered when trying to implement the parallel procedure:

i) How to generate random functions, not necessarily multigaussian, with a given covariance function? Various techniques can be found in the literature (spectral method, turning bands method, dilution, tessellation ...). Each technique has a range of validity and an efficiency that has to be determined. Special emphasis will be given to the comparison of the turning bands method with the spectral method from the standpoint of efficiency.

ii) What is the number n of random functions that should be generated? The answer to this question is not very easy because a deviation from multigaussianity can result either from a convergence problem (if n is not large enough) or from a support problem (if the simulation domain is not large w.r.t. the covariance range). Both problems can be addressed by considering expansions related to the Central Limit Theorem, in particular the Berry-Esséen Theorem (Feller, 1971).

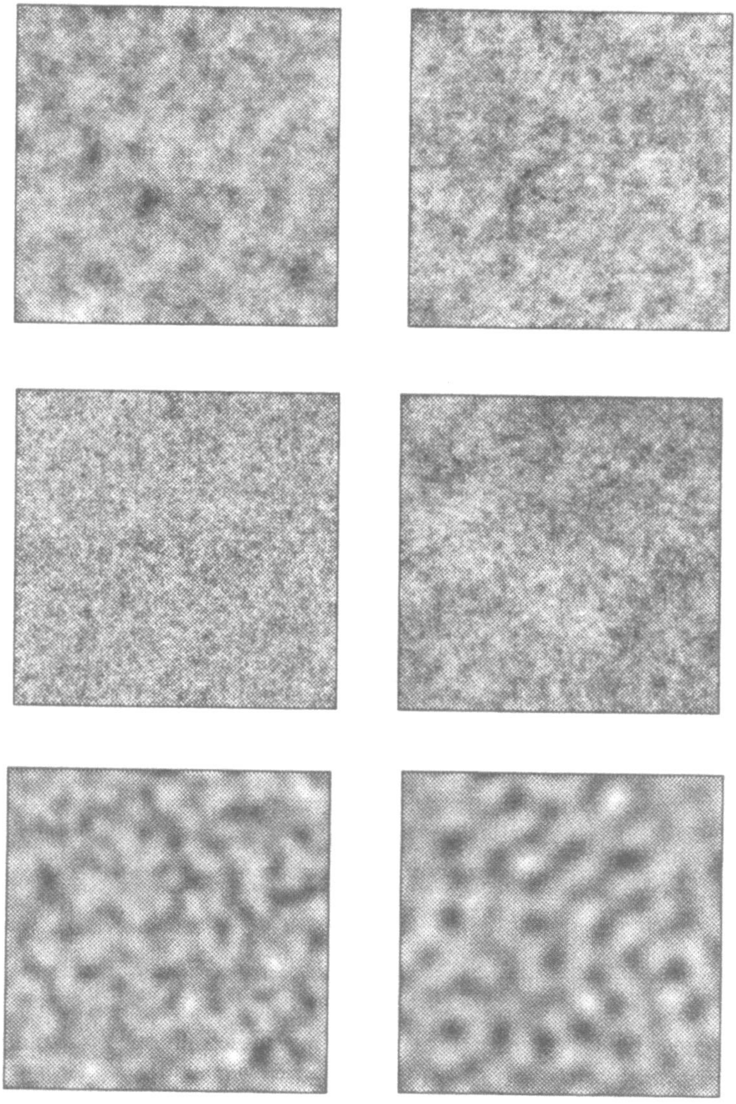

Figure 1: Some realizations of standard multigaussian random functions with different covariance functions. From top to bottom and left to right, spherical, exponential, stable, hyperbolic, Gaussian and cardinal sine covariances.

2 Some examples of MGRF

The following table gives the covariance models that will be considered throughout this paper.

Spherical	$C(h) = \left(1 - \frac{3}{2}\frac{\lvert h\rvert}{a} - \frac{1}{2}\frac{\lvert h\rvert^3}{a^3}\right) 1_{\lvert h\rvert \leq a}$
Exponential	$C(h) = \exp\left\{-\frac{\lvert h\rvert}{a}\right\}$
Stable	$C(h) = \exp\left\{-\sqrt{\frac{\lvert h\rvert}{a}}\right\}$
Hyperbolic	$C(h) = \frac{1}{1 + \frac{\lvert h\rvert}{a}}$
Gaussian	$C(h) = \exp\left\{-\frac{\lvert h\rvert^2}{a^2}\right\}$
Cardinal sine	$C(h) = \frac{\sin\left\{\frac{\lvert h\rvert}{a}\right\}}{\frac{\lvert h\rvert}{a}}$

These covariances have been chosen because they possess features (namely their behaviour at the origin, at the vicinity of the range, or at large scales) which are specific and different for each of them. These differences are visible on the realizations displayed in Figure 1.

How can realizations derived from different covariances be compared? The scale factor a of the spherical, exponential, stable and Gaussian covariances has been chosen so that the realizations behave similarly at large scale. This can be done by prescribing these four covariances to have the same integral range A_2 in two dimensions (Yaglom, 1986; Matheron, 1988; Lantuéjoul, 1991)

$$A_2 = \int_{\mathbb{R}^2} C(h)\, dh$$

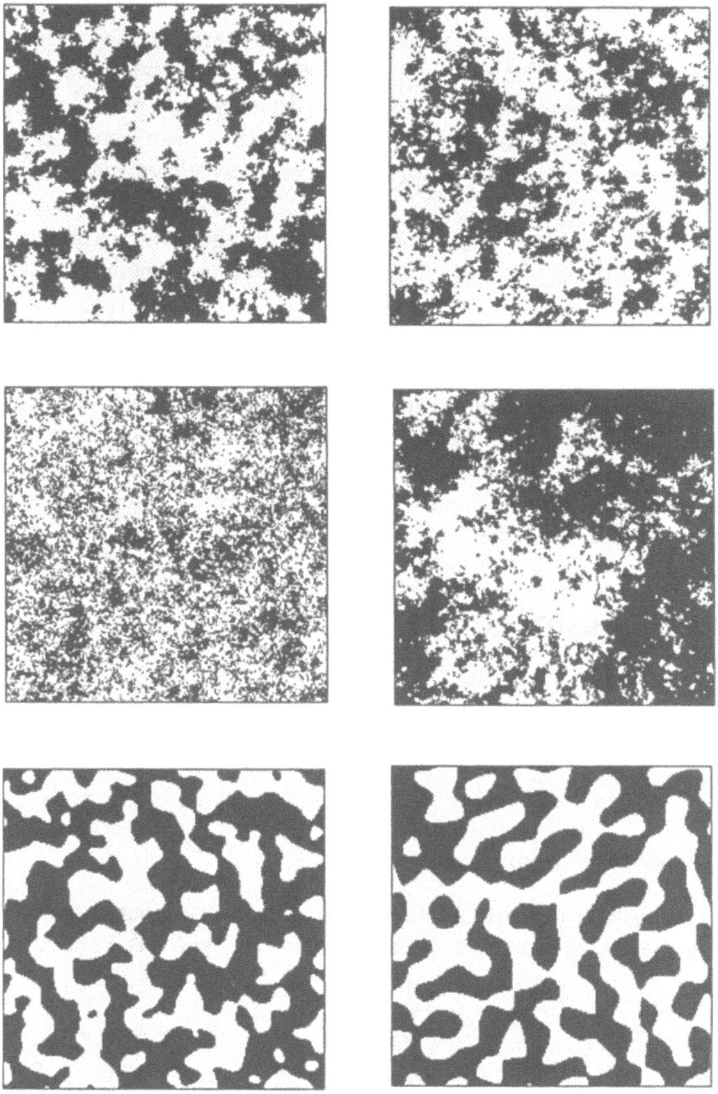

Figure 2: Thresholding the realizations of Figure 1 at cut-off 0.

On Figure 1, the simulation domain is actually 160 times larger than the integral range. Such a prescription is not possible for the hyperbolic covariance which has an infinite integral range. The consequence of an infinite integral range is that the realization does not present any sign of spatial homogeneity whatever the scale of observation. In the hyperbolic case, the scale factor has been assigned a value that gives the covariance the same behaviour at the origin as the exponential covariance.

The stable covariance has an infinite slope at the origin. This explains the broken aspect of the realization in Figure 1. The broken aspect remains visible after thresholding (cf. Figure 2).

Similarly, it is possible to choose the scale factors of the gaussian covariance and the cardinal sine covariance to give them the same smooth behaviour at the origin. The sine covariance is pseudoperiodic (but not strictly periodic: isotropy and periodicity are not compatible in two and three dimensions), which leads to realizations with a dendritic pattern (see Figures 1 and 2). The limbs of the dendrites have a width closely related to the scale factor.

3 Some basic methods

In this section, four basic methods are presented to simulate the Y_i's, namely the spectral method, the dilution method, the tessellation method and the turning bands method. An illustration of these four methods is given for the exponential covariance (see Figure 3).

3.1 The spectral method

A covariance is a positive definite function. If it is also continuous, Bochner's theorem states that it is the Fourier transform of a positive measure (the spectral measure), say χ

$$C(h) = \int_{\mathbb{R}^3} e^{i < u, h >} d\chi(u)$$

Moreover, since $C(0) = 1$, χ is a probability distribution. It is not difficult to show that if Ω is a random vector with distribution χ, and if Φ is a variable with a uniform distribution over $[0, 2\pi[$ independent of Ω, then the random function defined as

$$Y(x) = \sqrt{2}\cos(< \Omega, x > +\Phi)$$

has C as its covariance function. The spectral method, developed by Shinozuka (1972) among others, is quite general and can be easily implemented. For example,

Figure 3: Examples of random functions that can be used to simulate a MGRF with an exponential covariance. From top to bottom and left to right, the spectral method, the dilution method, the tessellation method and the turning bands method.

the spectral measure of the exponential covariance has the density

$$f(u) = \frac{1}{\pi^2} \frac{a^3}{(1+a^2|u|^2)^2}$$

The simulation of the density f can be achieved either by considering f as a gamma mixture of gaussian distributions

$$f(u) = \int_0^{+\infty} \frac{e^{-t}}{\sqrt{\pi t}} \left(\frac{2a^2t}{2\pi}\right)^{\frac{3}{2}} e^{-\frac{2a^2t|u|^2}{2}} dt$$

or by an acceptance-rejection method.

3.2 The dilution method

In contrast to the spectral method, the dilution method is not general. It is applicable only in the case where the covariance can be expressed as

$$C(h) = \theta \int_{\mathbb{R}^3} g(x)g(x+h)dx$$

A dilution random function (Matérn, 1986; Serra, 1968) is a moving average of functions called primary functions located at random points. More formally

$$Y(x) = \sum_{u \in \mathcal{P}} \varepsilon(u)\, g(x-u)$$

where $\mathcal{P}$ is a Poisson point process in $\mathbb{R}^3$ of intensity θ, ε is a family of independent standard random variables and g is a numerical function. In the exponential case, the primary function is

$$g(x) = \frac{1}{\sqrt{2\pi a}} \frac{\exp\left\{-\frac{|x|}{a}\right\}}{\frac{|x|}{a}}$$

In the present case, the family ε has been introduced to ensure that Y has a 0 mean value. In a more general setting, it is possible to replace the family εg by a family of independent random functions, which is not significantly different from the random tokens method developed by Alfaro (1979).

From a practical point of view, there are other limitations. A proper numerical implementation requires the primary function g to be bounded and to have a bounded support. This is the reason why the dilution method is quite suitable for simulating a spherical covariance function, but is not recommended for an exponential one.

3.3 The tessellation method

In this method, the whole space is divided into a stationary population of random cells. The cells are then assigned independent standard random values.

There exist many models to partition the space: the Poisson polyhedra (Matheron, 1971), the Voronoï polyhedra (Miles, 1972) or the dead leaves cells (Matheron, 1968a; Jeulin, 1991), just to name a few. For instance, in the exponential case, the cells are Poisson polyhedra (Matheron, 1971).

Nevertheless the tessellation method is not general, as the generated covariance coincides with the geometric covariogram of the cells. The degree of generality of the tessellation method is still unknown.

3.4 The turning bands method

Originated by Matheron (1972, 1973), the turning bands method is a stereological device designed to reduce a multidimensional simulation into unidimensional ones.

Let C_3 be the polar form of C. In other words, C_3 verifies

$$C(h) = C_3(|h|)$$

The turning bands methods shows that it is sufficient to simulate a stationary unidimensional random function X with covariance

$$C_1(r) = \frac{d}{dr}(r\ C_3(r))$$

X is then spread throughout the space

$$Y(x) = X(<\Theta, x>)$$

where Θ is a unit vector with a uniform direction. The choice of the manner of simulating X is totally free.

Consider, for instance, the case of an exponential covariance function (cf. Figure 4). Its polar form is

$$C_3(r) = \exp\left\{-\frac{r}{a}\right\}$$

which leads to

$$C_1(r) = \left(1 - \frac{r}{a}\right) \exp\left\{-\frac{r}{a}\right\}$$

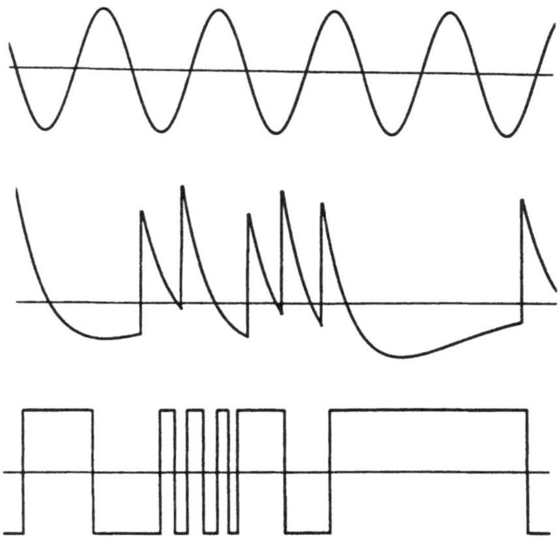

Figure 4: Simulation of an exponential covariance using the turning bands method. Three models for the unidimensional simulation. From top to bottom, the spectral method, the dilution method and the migration method.

The simulation of X can be done by dilution with the primary function

$$g(t) = \left(1 - \frac{t}{a}\right) \exp\left\{-\frac{t}{a}\right\} 1_{t \geq 0}$$

(Matheron, 1972; Journel, 1978), but this is not fully satisfactory as g does not have a bounded support. A possible alternative that does not require any truncation is to simulate X using the spectral method. A third possibility that has been inspired by migration techniques (Matheron, 1968b) also exists:
i) generate a Poisson point process that partitions $\mathbb{R}$ into independent exponential intervals of mean length $2a$.
ii) split each interval into two halves, and set the first half to $+1$, the second half to -1.

Consider the spherical model as a second example (cf Figure 5).

$$C_1(r) = \left(1 - 3\frac{r}{a} - 2\frac{r^3}{a^3}\right) 1_{r \leq a}$$

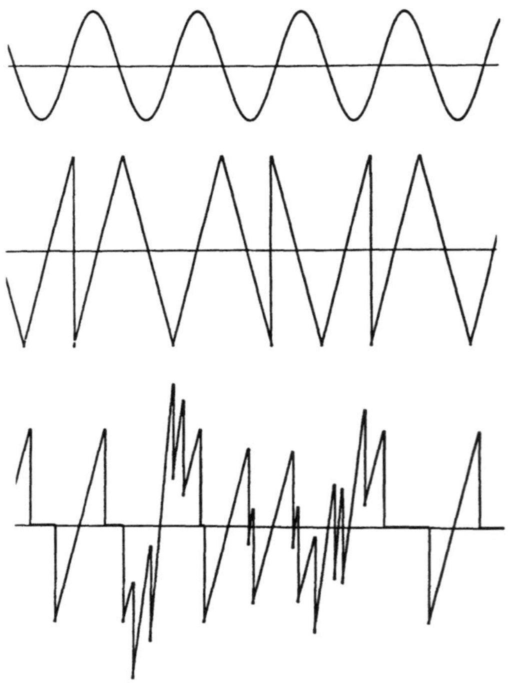

Figure 5: Simulation of a spherical covariance using the turning bands method. Three models for the unidimensional simulation. From top to bottom, the spectral method, the partition method and the dilution method.

In this case, X can be simulated using the spectral method (but the implementation is not easy due to the complicated shape of the spectral density), or using the dilution method with the primary function

$$g(t) = t\, 1_{|t| \leq a}$$

(Matheron, 1972, Journel, 1978). In contrast to the exponential case, the dilution method is here numerically quite suitable. Of course, other possibilities exist, such as that is described below (called here a partition method)
i) partition $\mathbb{R}$ into intervals of length a.
ii) within each interval, generate a linear function ranging from $-\sqrt{3}$ to $\sqrt{3}$. The function has an equal probability of being increasing or decreasing. The functions associated with different intervals are independent.

Now the question is to decide which amongst all the possible simulation methods is the best one.

4 Comparison of the spectral method with the turning bands method

Let us start by dispelling a common misconception. A priori, one could think that the spectral method is inferior because so few random deviates have to be generated to get a simulation. Such an idea suggests the following question: how much information is necessary to produce a realization of a stationary multigaussian random function with a given covariance in a given simulation domain? It turns out that this question has a very simple answer. A general simulation theorem states that any simulation on a finite space can be done using one and only one uniform value (Bouleau, 1986). The intuitive reason is that two independent uniform values can be derived from one uniform value (an explicit construction given in Bouleau is based upon the dyadic expansion of the uniform value). Of course, this result is only of theoretical interest, but it clearly shows that what is important is not the amount of information that has been used for the simulation, but the way in which this information has been used.

The random functions produced by the spectral method can be written $Y(x) = X(<\Theta, x>)$, where X is a unidimensional random function defined by $X(t) = \sqrt{2}\cos(Rt + \Phi)$, and where (R, Θ) are the polar coordinates of the vector Ω. This means that the spectral method, exactly as the turning bands method, is a spreading method. However, there is a notable difference. In the turning bands method, the unidimensional random function X can be chosen to be ergodic, in the sense that its spatial distribution can be determined from any of its realizations. In the spectral method, X is not in general ergodic.

Consider two simulations, the first one obtained using the spectral method

$$Y_s^{(n)}(x) = \frac{\sqrt{2}}{\sqrt{n}} \sum_{i=1}^{n} \cos(< \Omega_i, x > + \Phi_i)$$

the second one using the turning bands method

$$Y_t^{(n)}(x) = \frac{1}{\sqrt{n}} \sum_{i=1}^{n} Y_i(< \Theta_i, x >)$$

The covariances of the two simulations are respectively

$$C_s^{(n)}(h) = \frac{1}{n} \sum_{i=1}^{n} \cos(< \Omega_i, h >)$$

and

$$C_t^{(n)}(h) = \frac{1}{n} \sum_{i=1}^{n} C_1(< \Theta_i, h >)$$

As the Θ_i and the Ω_i are random, the two covariances C_s and C_t are actually random. If the directions of the Ω_i are the Θ_i, then it can be shown that

$$E\left\{C_s^{(n)}(h) \mid \Theta_1, ..., \Theta_n\right\} = C_t^{(n)}(h)$$

(cf. Appendix), which implies

$$E\left\{C_s^{(n)}(h) \mid C_t^{(n)}(h)\right\} = E\left\{C_s^{(n)}(h) \mid \Theta_1, ..., \Theta_n \mid C_t^{(n)}(h)\right\} = C_t^{(n)}(h)$$

This is Cartier's formula (Matheron, 1984), which shows that the distribution of $C_s^{(n)}(h)$ is more dispersed than that of $C_t^{(n)}(h)$. From Cartier's formula, it can be derived that

$$E\left\{\phi\left[C_s^{(n)}(h)\right]\right\} \geq E\left\{\phi\left[C_t^{(n)}(h)\right]\right\}$$

holds for any numerical convex function ϕ. This gives for instance

$$Var\left\{C_s^{(n)}(h)\right\} \geq Var\left\{C_t^{(n)}(h)\right\}$$

in the particular case where $\phi(t) = t^2 - C^2(h)$. The difference of dispersion between both covariances shows that the turning bands method applied to a unidimensional ergodic function is the spreading method that possesses the fastest rate of convergence from the standpoint of the covariance.

An illustration of this result is given hereunder. A way to prove that the hyperbolic function

$$C(h) = \frac{1}{1 + |h|}$$

is a covariance is to write it as as a positive linear combination of exponential covariances

$$C(h) = \int_0^{+\infty} e^{-t} \, e^{-t|h|} \, dt$$

This decomposition suggests the following simulation procedure using the turning bands method. This consists of proceeding as if one wanted to simulate an exponential covariance. The only difference is that the scale factor is random and must be generated independently from one line to another. Unfortunately the random functions simulated this way along each line are not ergodic, and the procedure therefore cannot be optimal. It is much better to observe that

$$C_1(r) = \frac{1}{(1+r)^2}$$

is convex, which enables ergodic simulations along the lines using the tessellation method (Matheron, 1987). Here the cells are random intervals with length distribution

$$l(x) = \frac{3}{(1+x)^4}$$

Note in passing, that to have C_1 convex is rather exceptional. Usually C_3 vanishes rapidly at infinity, so

$$\int_0^{+\infty} C_1(r) \, dr = [r \, C_3(r)]_0^{+\infty} = 0$$

and therefore C_1 cannot be convex.

5 How many random functions?

Suppose n has been fixed. Is the spatial distribution of

$$Y^{(n)} = \frac{Y_1 + \cdots + Y_n}{\sqrt{n}}$$

close to a standard multigaussian distribution? Equivalently, if $x_1, ..., x_p$ stand for any set of points, and if $\lambda_1, ..., \lambda_p$ denote any set of numbers, how does the distribution of $\sum_{j=1}^{p} \lambda_j Y_1^{(n)}(x_j)$ depart from a gaussian distribution with mean value 0 and variance $\sum_{j,k=1}^{p} \lambda_j \lambda_k C(x_j - x_k)$?

To address this problem, two different approaches are considered in this paper, namely the distribution approach and the moment approach.

5.1 The distribution approach

Let $U_1, ..., U_n, ...$ be a sequence of independent and identically distributed random variables. According to the strong Law of the Large Numbers, the average $(U_1 + \cdots + U_n)/n$ converges almost surely towards the mean value m of the distribution, supposed to be finite

$$\lim_{n \longrightarrow +\infty} \frac{U_1 + \cdots + U_n}{n} = m \quad a.s.$$

In the case where the variance σ^2 is also finite, the Central Limit Theorem states that the distribution of the average tends to be gaussian around the mean value:

$$\lim_{n \longrightarrow +\infty} P\left\{ \frac{\frac{U_1 + \cdots + U_n}{n} - m}{\frac{\sigma}{\sqrt{n}}} < u \right\} = G(u)$$

where G denotes the standard gaussian distribution function. Now the question is to know for what value of n onwards the average can be considered as gaussian. Provided that the distribution of the variables admits a finite third order absolute moment $\mu_3 = E\{|U_i - m|^3\}$, the answer is given by the Berry-Esséen theorem (Feller, 1971; Rao, 1984). This theorem gives an upper bound for the difference between the standardized distribution of the average, and the standard gaussian distribution

$$\sup_{u \in \mathbf{R}} \left| P\left\{ \frac{\frac{U_1 + \cdots + U_n}{n} - m}{\frac{\sigma}{\sqrt{n}}} < u \right\} - G(u) \right| < \frac{\alpha}{\sqrt{n}} \frac{\mu_3}{\sigma^3}$$

where α is a numerical constant less than 1.32132.

If this approach is applied to the sequence of random variables

$$U_i = \sum_{j=1}^{p} \lambda_j Y_i(x_j) \qquad i = 1, 2, ...$$

one obtains

$$\sup_{u \in \mathbf{R}} \left| P\left\{ \frac{\sum_{j=1}^{p} \lambda_j Y^{(n)}(x_j)}{\sigma} < u \right\} - G(u) \right| < \frac{\alpha}{\sqrt{n}} \frac{\mu_3}{\sigma^3}$$

The variance of the U_i's depends only on the covariance of the Y_i's

$$\sigma^2 = \sum_{j,k=1,p}^{p} \lambda_j \lambda_k \, C(x_j - x_k)$$

but the third order absolute moment

$$\mu_3 = E\left\{\left|\sum_{j=1}^{p}\lambda_j Y_i(x_j)\right|^3\right\}$$

involves the whole multivariate distribution of $(Y_i(x_1), ..., Y_i(x_p))$. For this reason, it is not always easily tractable. A way to bypass the calculation of μ_3 is to apply the Cauchy-Schwarz inequality

$$\mu_3 \leq \sigma\ \sqrt{\mu_4}$$

where μ_4 is the fourth order moment

$$\mu_4 = E\left\{\left[\sum_{j=1}^{p}\lambda_j Y_i(x_j)\right]^4\right\}$$

so that

$$\sup_{u\in\mathbf{R}}\left|P\left\{\frac{\sum_{j=1}^{p}\lambda_j Y^{(n)}(x_j)}{\sigma} < u\right\} - G(u)\right| < \frac{\alpha}{\sqrt{n}}\frac{\sqrt{\mu_4}}{\sigma^2}$$

This of course does not constitute an improvement of the Berry-Esséen inequality, but the right hand side member can be calculated.

As an illustration, consider the simulation of an exponential covariance using the turning bands method (migration technique). The difference

$$U_i = Y_i(x) - Y_i(x+h)$$

gives

$$\sigma^2 = 2(1-\rho) \qquad \mu_4 = 8(1-\rho)$$

with $\rho = \exp\{-|h|/a\}$. The formula derived from the Berry-Esséen inequality becomes

$$\sup_{u\in\mathbf{R}}\left|P\left\{\frac{Y^{(n)}(x) - Y^{(n)}(x+h)}{\sqrt{2(1-\rho)}} < u\right\} - G(u)\right| < \frac{\alpha}{\sqrt{n}}\sqrt{\frac{2}{1-\rho}}$$

The right hand side member is a monotonic decreasing function of $|h|$, and is, in particular, very large for small $|h|$ values. Supposing $|h| \geq a_0$ and noting $\rho_0 = \exp\{-a_0/a\}$, one has

$$\frac{\alpha}{\sqrt{n}}\sqrt{\frac{2}{1-\rho}} < \varepsilon \quad \Longleftrightarrow \quad n \geq n_0 = \frac{2\alpha^2}{\varepsilon^2(1-\rho_0)}$$

For $a_0 = a/10$ and $\varepsilon = 0.1$, this gives $n_0 \approx 3670...$

If the simulation is carried out using the spectral method instead, then one gets $\mu_4 = 3(1-\rho)(3-\rho)$, which leads to $n_0 \approx 5770$.

These values obtained for n_0 are of course unrealistically large and are not confirmed in practice. Indeed they express several facts:

i) the criterion under study, i.e. the *distribution* of $Y^{(n)}(x) - Y^{(n)}(x+h)$, is quite drastic. In next subsection, a less severe criterion will be presented.

ii) the inequality derived from the Berry-Esséen theorem is rather loose. For n fixed, the right hand side member of the equation becomes larger than 1 as $|h|$ is close to 0.

iii) the power $\frac{1}{2}$ that affects the factor n stems from the independence of the Y_i, which implies not only the independence of the X_i, but also the independence of the line directions. In a more general presentation of the turning bands method, taking independent directions is not necessary (Matheron, 1971), and this is certainly not the best possible choice. It is preferrable to use as regular a set of directions as possible (see section 7) to ensure a faster rate of convergence (Mantoglou and Wilson, 1982). However, the gain in the rate of convergence is not known at present.

iv) things go specially bad with the difference $Y^{(n)}(x) - Y^{(n)}(x+h)$. If the sum $Y^{(n)}(x) + Y^{(n)}(x+h)$ had been considered instead, then one would have obtained $n_0 \approx 180$.

5.2 The moment approach

The moment approach consists of comparing the moments of $Y^{(n)}$ with those of a standard multigaussian random function. As all of the odd moments are equal to 0, and as the second moments are fully specified by the covariance, the simplest non trivial moments to consider are the fourth order ones.

It can be shown that

$$E\left\{\left[\sum_{j=1}^{p} \lambda_j Y^{(n)}(x_j)\right]^4\right\} - 3\sigma^4 = \frac{1}{n}\left(\mu_4 - 3\sigma^4\right)$$

where σ^2 and μ_4 have been defined earlier. It should be noted that $3\sigma^4$ is just the fourth order moment of a gaussian variable with variance σ^2.

In the moment approach, n is chosen so that the relative discrepancy between the fourth order moments is less than a critical value, say ε. This gives

$$\frac{1}{n}\left|\frac{\mu_4 - 3\sigma^4}{3\sigma^4}\right| < \varepsilon$$

This procedure has been applied to the difference $Y^{(n)}(x) - Y^{(n)}(x+h)$ (which yields a relative variance for the variogram) using the same simulation techniques. Keeping $\varepsilon = 0.1$, one obtains $n_0 \approx 60$ for the turning bands method and $n_0 \approx 45$ for the spectral method. Clearly, relaxing the criterion has dramatic consequences.

6 On statistical fluctuations

The fact that simulations are done not in the whole space $\mathbb{R}^3$, but in a limited domain D has several implications from a variographic standpoint.

Let Y be a stationary random function with variogram γ. By definition, the (probabilistic version of the) regional variogram of Y in D is

$$\Gamma_D(h) = \frac{1}{2\,|D \cap D_h|} \int_{D \cap D_h} [Y(x+h) - Y(x)]^2 \, dx$$

(Matheron, 1988). The random variable $\Gamma_D(h)$ is an unbiased estimate of $\gamma(h)$

$$E\{\Gamma_D(h)\} = \gamma(h)$$

and its variance is

$$Var\{\Gamma_D(h)\} = \frac{1}{2\,|D \cap D_h|^2} \int\int [\gamma(x-y+h) + \gamma(x-y-h) - 2\gamma(x-y)]^2 \, dx \, dy$$

in the multigaussian case (the integration domain for both integrals is $D \cap D_h$). $Var\{\Gamma_D\}$ is negligible if and only if the dimensions of D are very large w.r.t. the range of γ.

Now let $\Gamma_D^{(n)}$ be the regional variogram of $Y^{(n)}$ in D. It can be shown that the variance of $\Gamma_D^{(n)}$ is a weighted average of the variance of Γ_D and of the variance of the regional variogram $\Gamma_{D,i}$ of the Y_i's

$$Var\{\Gamma_D^{(n)}(h)\} = \left(1 - \frac{1}{n}\right) Var\{\Gamma_D(h)\} + \frac{1}{n} Var\{\Gamma_{D,i}(h)\}$$

If the size of D is not very large w.r.t. the range of γ, then $Var\{\Gamma_D^{(n)}\}$ does not vanish. This means that $\Gamma_D^{(n)}$ has statistical fluctuations. The fluctuations of $\Gamma_D^{(n)}$ and Γ_D tend to have the same order of magnitude as n becomes very large. Conversely, unless it has been designed to respect the constraint of a regional variogram, a simulation procedure which produces regional variograms in a limited domain without statistical fluctuations is necessarily erroneous.

7 Generating lines with 'regular' directions

Suppose that we want to calculate the integral of a numerical function f defined on a 100-dimensional unit cube. The standard procedure, which consists of computing the mean value of f over a regular grid of points, does not work here. For instance, even a rough discretization of the cube, such as 3 points along each coordinate axis, would require 3^{100} values to be calculated, which is far beyond the capabilities of standard computers. Another approach, the Monte-Carlo approach, is to take the mean value of f over a set of independent and uniform points in the unit cube. This approach is advantageous in that it avoids calculating a large number of f values. However, it is not totally satisfactory, for the spatial distribution of independent and uniform points is far from homogeneous. Some clusters of points can be observed from place to place, whereas some other parts of the unit cube are totally empty.

A third approach is to compute the mean value of f over an equidistributed sequence of points. A sequence of points $x_1, ..., x_n, ...$ is said to be equidistributed if, for any B contained in the unit cube, the proportion of points falling within B from among the first n points tends to the volume of B as n becomes very large.

As a matter of fact, a realization of a sequence of independent uniform points is equidistributed. But the trick is that there exist sequences that converge much faster than uniform points (Bouleau, 1988). The rule of construction usually relies on the intuitive principle that the nth point of the sequence must be located in the unit cube as far as possible from $x_1, ..., x_{n-1}$.

Freulon (1991,1992) considered a sequence of non uniform equidistributed points to generate a set of directions on the unit sphere. His algorithm is the following: he considers the binary and the ternary expansions of any integer $n = 1, 2, ...$

$$n = a_0 + 2a_1 + \cdots + 2^p a_p \qquad a_i = 0, 1$$

$$n = b_0 + 3b_1 + \cdots + 3^q b_q \qquad b_j = 0, 1, 2$$

from which 2 numbers between 0 and 1 are generated

$$u_n = \frac{a_0}{2} + \frac{a_1}{4} + \cdots + \frac{a_p}{2^{p+1}}$$

$$v_n = \frac{b_0}{3} + \frac{b_1}{9} + \cdots + \frac{b_q}{3^{q+1}}$$

The coordinates of the nth point of the sequence are

$$x_n = \left(\cos(2\pi u_n)\sqrt{1 - v_n^2}\ ,\ \sin(2\pi u_n)\sqrt{1 - v_n^2}\ ,\ v_n\right)$$

Figure 6 shows a comparison of 400 independent and uniform points with the 400 points generated according the previous algorithm.

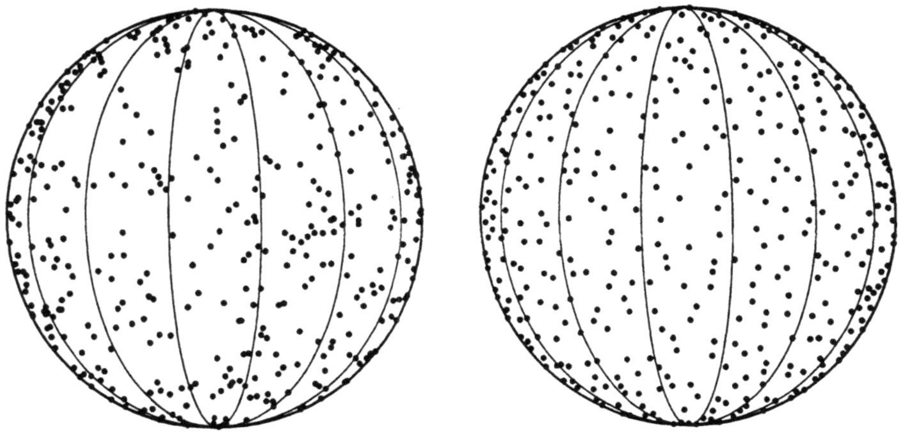

Figure 6: Generation of 400 points on the unit sphere. On the left, the points are independent and uniform. On the right, they are located according to an equidistributed sequence.

Of course, the sequence can be considered as defined up to a random rotation. The rotation is used to avoid building two different simulations using the same set of directions.

8 Conclusions

In this paper, a parallel method has been presented to simulate multigaussian random functions. This method includes the turning bands method and its particular case, the spectral method. It has been shown that the turning bands method can be implemented in a continuous way along the lines, without any discretization or truncature problem.

Several tools have been given to assess the number of basic functions required for the simulation. This number depends upon the type of covariance function, and also upon the algorithm that has been implemented. Consequently, no magic number can be recommended, but more than 15 basic random functions are undoubtedly necessary. In any case, isn't it amazing to recommend 180 lines for a two-dimensional simulation but only 15 for a three-dimensional one? It is true that at the time when the first algorithms where designed (Journel, 1978), display procedures were not so efficient, and one was content with a relatively good respect of the variogram model. Any departure was usually imputed to statistical fluctuations which are encountered as soon as the simulation field is large with respect of the range of the covariance.

As for the turning bands method is concerned, it is better to use a relatively regular set of directions instead of a set of independent and uniform directions, but the actual gain is not known at present.

In Appendix 2, a program is given to allow the interested readers to gain experience with the parallel simulation methods. This program has been written for the isotropic case. However, only minor changes are required to handle classical anisotropies (geometric or zonal).

Acknowlegments: The author is grateful to Dr. Hari Pandalai from I.I.T. Bombay for carefully reading the first draft of this paper.

References

Alabert, F.G. (1987) "Stochastic imaging of spatial distributions using hard and soft data", Master's thesis, Stanford University.

Alfaro Sironvalle, M. (1979) "Etude de la robustesse des simulations de fonctions aléatoires", Geostatistics Doctoral Thesis, School of Mines of Paris.

Bouleau, N. (1986) Probabilités de L'ingénieur, Hermann, Paris.

Feller, W. (1971) An introduction to Probability Theory and its Applications, Vol. 2, Wiley, New-York.

Freulon, X. and de Fouquet, C. (1991) "Remarques sur la pratique des bandes tournantes à 3 dimensions" , Cahiers de Géostatistique N°1, pp. 101,117.

Freulon, X. (1992) "Conditionnement du modèle gaussien par des inégalités ou des randomisées", Geostatistics Doctoral Thesis, School of Mines of Paris.

Jeulin, D. (1991) "Modèles morphologiques de structures aléatoires et de changement d'échelle", Doctoral Thesis, University of Caen.

Journel, A.G. and Huijbregts, C.J. (1978) Mining Geostatistics, Academic Press, London.

Lantuéjoul, Ch. (1991) "Ergodicity and Integral Range", Journal of Microscopy, Vol. 161-3, pp. 387-404.

Mantoglou, A. and Wilson, J.L. (1982) "The Turning Bands Method for Simulation of Random Fields Using Line Generation by A Spectral Method", Water Res. Res., Vol. 18-5, pp. 1379-1394.

Matérn, B. (1986) Spatial Variation, Lecture Notes in Statistics Vol. 36, Springer-Verlag, Berlin (2nd edition).

Matheron, G. (1968a) "Schéma Booléen séquentiel de partition aléatoire", Internal Report

N-83, Centre de Morphologie Mathématique, Fontainebleau.

Matheron, G. (1968b) "Processus de renouvellement et pseudo-périodicités", Internal Report N-71, Centre de Morphologie Mathématique, Fontainebleau.

Matheron, G. (1971) "Les polyèdres Poissonniens isotropes", 3ème Symposium Européen sur la Fragmentation, Cannes, 5-8 octobre 1971.

Matheron, G. (1973) "The intrinsic random functions and their applications", Adv. Appl. Prob., Vol. 5, pp. 439-468.

Matheron, G. (1984) The selectivity of the distributions and "the second principle of geostatistics". In VERLY, G. et al. (eds.) *Geostatistics for natural ressources characterization*, pp. 421-423, NATO ASI Series C, Vol 122, Reidel, Dordrecht.

Matheron, G. (1987) "Simulation de fonctions aléatoires admettant un variogramme concave donné", Internal Report N-11/87, Centre de Géostatistique, Fontainebleau.

Matheron, G. (1988) Estimating and Choosing, Springer-Verlag, Berlin.

Miles, R.E. (1972) "The random division of space", special supplement to Adv. in Applied Prob., pp. 243-266.

Rao, M.M. (1984) Probability Theory with Applications, Academic Press, Orlando.

Serra, J. (1968) "Fonctions aléatoires de Dilution", Internal Report C-12, Centre de Morphologie Mathématique, Fontainebleau.

Shinozuka, M. and Jan, C.M. (1972) "Digital simulation of random processes and its applications", J. of Sound and Vib., Vol. 25-1, pp. 111-128.

Yaglom, A.M. (1986) Correlation Theory of Stationary and related Random Functions, Springer Series in Statistics, New York.

Appendix 1: Comparison of the spectral method with the turning bands method.

For the sake of simplicity, it is assumed here that the covariance C vanishes rapidly at infinity. Consequently its spectral measure admits a density denoted f

$$C(h) = \int_{\mathbb{R}^3} e^{\,i < u,h >} f(u)\, du$$

In polar coordinates, this gives

$$C_3(r) = 4\pi \int_0^{+\infty} \frac{\sin(rt)}{r}\, t\, f_3(t)\, dt$$

with $f(u) = f_3(|u|)$. The formula for the covariance can be inverted

$$f_3(t) = \frac{1}{2\pi^2} \int_0^{+\infty} \frac{\sin(tr)}{t}\, r\, C_3(r)\, dr$$

Similarly, let f_1 the spectral density of C_1

$$C_1(r) = \int_{\mathbb{R}} \cos(rt)\, f_1(|t|)\, dt$$

or equivalently

$$f_1(t) = \frac{1}{\pi} \int_0^{+\infty} \cos(tr)\, C_1(r)\, dt$$

Replacing C_1 by its expression

$$C_1(r) = \frac{d}{dr}\,(rC_3(r))$$

and integrating by parts, one finds

$$f_1(t) = \frac{1}{\pi} \int_0^{+\infty} t\, \sin(tr)\, r\, C_3(r)\, dr$$

that is

$$f_1(t) = 2\pi t^2 f_3(t)$$

Let Ω a random vector with distribution $f(u) = f_3(|u|)$. In polar coordinates, Ω can be written (R, Θ), and the modulus R follows the distribution $4\pi t^2 f_3(t)$. Then

$$\begin{aligned} E\{\cos(< \Omega, h >) \mid \Theta\} &= E\{\cos(R < \Theta, h >) \mid \Theta\} \\ &= 4\pi \int_0^{+\infty} \cos(t < \Theta, h >)\, t^2\, f_3(t)\, dt \\ &= 2 \int_0^{+\infty} \cos(t < \Theta, h >)\, f_1(t)\, dt \\ &= C_1(< \Theta, h >) \end{aligned}$$

Appendix 2: A simulation program.

The aim of this program is to give people the possibility to simulate stationary isotropic multigaussien random functions with the covariance functions given in Table 1. This program has basically two input arguments, that is the simulation field (a three-dimensional grid) and the model for the covariance function.
The source code has been written using several IMSL functions and subroutines:

- rnopt selects the uniform $(0,1)$ generator
- rnget retrieves the current value of the seed
- rnset initializes the seed used in the generator

- drnunf generates a uniform $(0,1)$ value
- drrnoa generates a standard gaussian value
- drngam generates a gamma value

The subroutine 'drngam' is actually used to simulate a gamma variable of parameter 0.5. An alternative way is to consider half the square of a standard gaussian variable. We recall here that if u and v are two independent uniform $(0,1)$ values, then

$$x = \sqrt{-2\ln u}\ \cos(2\pi v)$$

is standard gaussian distributed.

```
c
c
c  Non conditional simulation of three-dimensional stationary
c  isotropic multigaussian random functions
c
c
      parameter (ndmax=2000)
      parameter (nxmax=1000)
      parameter (ntmax=5000)
c
      implicit real*8 (a-h,o-z)
c
      integer struct,seed(ndmax)
c
      dimension cdx(ndmax),cdy(ndmax),cdz(ndmax)
      dimension buf(nxmax)
      dimension t(ntmax),tv(ntmax)
c
      character*30 filout
      character cr
c
      common iseed
c
      data pi/3.14159265d0/
      data cr/13/
c
c  formats
c
2000  format (3i5,6f12.6)
2100  format (10f12.6)
3000  format ('iz,iy=',2i5,$)
3100  format (a1,'iz,iy=',2i5,$)
3200  format (a1)
c
c
c---------------------------------------------------------------
c
c  characteristics of the simulation
c
c
c  geometric description of the simulation field
c
      write (6,*)'simulation field'
      write (6,*)'----------------'
      write (6,*)'number of points along the x, y and z axes?'
      read (5,*)nx,ny,nz
      if (nx.gt.nxmax)stop   'too many points along the x axis'
      write (6,*)'minimal coordinates of the simulation field?'
      read (5,*)xmin,ymin,zmin
      write (6,*)'lags along the x, y and z axes?'
      read (5,*)echx,echy,echz
c
c  structure
c
      write (6,*)' '
      write (6,*)'variographic information '
      write (6,*)'-------------------------'
      write (6,*)'type of variogram?'
      write (6,*)'spherical         1'
      write (6,*)'exponential       2'
      write (6,*)'stable            3'
      write (6,*)'hyperbolic        4'
      write (6,*)'gaussian          5'
      write (6,*)'cardinal sine     6'
      read (5,*)struct
      write (6,*)'sill?'
      read (5,*)sigma2
      write (6,*)'scale factor?'
      read (5,*)a
c
c  about the random functions
c
      write (6,*)' '
      write (6,*)'Number of basic random functions?'
      write (6,*)'---------------------------------'
      read (5,*)nd
c
```

```
c  other informations
c
      write (6,*)' '
      write (6,*)'Name of the output simulation file?'
      write (6,*)'-----------------------------------'
      read (5,*)filout
      write (6,*)'random seed?'
      read (5,*)iseed
c
      open (unit=40,file=filout)
c
c-----------------------------------------------------------
c
c  initializations
c
      write (6,*)'initializations'
c
c  some features of the simulation field
c
      xmax=xmin+dfloat(nx-1)*echx
      ymax=ymin+dfloat(ny-1)*echy
      zmax=zmin+dfloat(nz-1)*echz
c
c  choose the uniform generator (IMSL procedure)
c
      call rnset(iseed)
      call rnopt(5)
c
c  normation factor
c
      sigma=dsqrt(sigma2/dfloat(nd))
      if (struct.eq.1)sigma=dsqrt(sigma2*3.d0/dfloat(nd))
      if (struct.eq.3)sigma=dsqrt(sigma2*2.d0/dfloat(nd))
      if (struct.eq.5)sigma=dsqrt(sigma2*2.d0/dfloat(nd))
      if (struct.eq.6)sigma=dsqrt(sigma2*2.d0/dfloat(nd))
c
c  initialize the minimum and the maximum of the simulation
c
      vmin=1.d+38
      vmax=-1.d+38
c
c  save some parameters
c
      write (40,2000)nx,ny,nz,xmin,ymin,zmin,echx,echy,echz
c
c-----------------------------------------------------------
c
c  prepare the simulation
c
      write (6,*)'preparation'
c
c  generate directions
c
      call vdc(nd,ndmax,cdx,cdy,cdz)
c
c  loop on the directions
c
      do id=1,nd
c
c  projection of the simulation field
c  onto a line of direction id
c
        t1=xmin*cdx(id)+ymin*cdy(id)+zmin*cdz(id)
        t2=xmin*cdx(id)+ymin*cdy(id)+zmax*cdz(id)
        t3=xmin*cdx(id)+ymax*cdy(id)+zmin*cdz(id)
        t4=xmin*cdx(id)+ymax*cdy(id)+zmax*cdz(id)
        t5=xmax*cdx(id)+ymin*cdy(id)+zmin*cdz(id)
        t6=xmax*cdx(id)+ymin*cdy(id)+zmax*cdz(id)
        t7=xmax*cdx(id)+ymax*cdy(id)+zmin*cdz(id)
        t8=xmax*cdx(id)+ymax*cdy(id)+zmax*cdz(id)
        tmin=min(t1,t2,t3,t4,t5,t6,t7,t8)
        tmax=max(t1,t2,t3,t4,t5,t6,t7,t8)
c
c  save the random seed (IMSL procedure)
c
        call rnget(seed(id))
c
c  spherical case
```

```
c
      if (struct.eq.1)then
        tdeb=tmin-a*drnunf()
        nt=ifix(sngl((tmax-tdeb)/a))+1
        if (nt.gt.ntmax)stop  'array ''t'' too small'
        do it=1,nt
          t(it)=dfloat(2*idnint(drnunf())-1)
        enddo
      endif
c
c  exponential case
c
      if (struct.eq.2)then
        b=2.d0*a
        t(1)=tmin+b*dlog(drnunf())
        t(2)=tmin-b*dlog(drnunf())
        nt=2
        do while (t(nt).le.tmax)
          nt=nt+1
          if (nt.gt.ntmax)stop  'array ''t'' too small'
          t(nt)=t(nt-1)-b*dlog(drnunf())
        enddo
      endif
c
c  stable case
c
      if (struct.eq.3)then
        call drnnoa(1,b)
        b=2.d0*a*b*b
        call drngam(1,0.5d0,c)
        call drnnoa(1,ux)
        call drnnoa(1,uy)
        call drnnoa(1,uz)
        r=dsqrt((ux*ux+uy*uy+uz*uz)/2.d0/c)/b
        phi=2.d0*pi*drnunf()
      endif

c
c  hyperbolic case
c
      if (struct.eq.4)then
        u=1.
        do while (u.gt.drnunf())
          u=dsqrt(drnunf())
        enddo
        t(1)=tmin-a*drnunf()*(1.d0/u-1.d0)
        tv(1)=dfloat(2*idnint(drnunf())-1)
        nt=1
        do while (t(nt).le.tmax)
          nt=nt+1
          if (nt.gt.ntmax)stop  'array ''t'' too small'
          t(nt)=t(nt-1)+a*(1.d0/drnunf()**0.3333-1.d0)
          tv(nt)=dfloat(2*idnint(drnunf())-1)
        enddo
      endif
c
c  gaussian case
c
      if (struct.eq.5)then
        call drnnoa(1,ux)
        call drnnoa(1,uy)
        call drnnoa(1,uz)
        r=dsqrt(2.d0*(ux*ux+uy*uy+uz*uz))/a
        phi=2.d0*pi*drnunf()
      endif
c
c  cardinal sine case
c
      if (struct.eq.6)then
        r=dfloat(2*idnint(drnunf())-1)/a
        phi=2.d0*pi*drnunf()
      endif
c
c  end of loop on the directions
c
    enddo
c
c-----------------------------------------------------------
```

```
c
c  simulation
c
      write (6,*)'simulation'
      write (6,3000)iz,iy
c
c  loops on the y and z coordinates
c
      do iz=1,nz
      do iy=1,ny
c
c  coordinates
c
        z=zmin+dfloat(iz-1)*echz
        y=ymin+dfloat(iy-1)*echy
c
c  reset buffer
c
        do ix=1,nx
          buf(ix)=0.d0
        enddo
c
c  loop on the directions
c
        do id=1,nd
c
c  projection of the simulation field
c  onto a line of direction id
c
          t1=xmin*cdx(id)+ymin*cdy(id)+zmin*cdz(id)
          t2=xmin*cdx(id)+ymin*cdy(id)+zmax*cdz(id)
          t3=xmin*cdx(id)+ymax*cdy(id)+zmin*cdz(id)
          t4=xmin*cdx(id)+ymax*cdy(id)+zmax*cdz(id)
          t5=xmax*cdx(id)+ymin*cdy(id)+zmin*cdz(id)
          t6=xmax*cdx(id)+ymin*cdy(id)+zmax*cdz(id)
          t7=xmax*cdx(id)+ymax*cdy(id)+zmin*cdz(id)
          t8=xmax*cdx(id)+ymax*cdy(id)+zmax*cdz(id)
          tmin=min(t1,t2,t3,t4,t5,t6,t7,t8)
          tmax=max(t1,t2,t3,t4,t5,t6,t7,t8)
c
c  restore the random seed (IMSL procedure)
c
          call rnset(seed(id))
c
c  simulation in the spherical case
c
          if (struct.eq.1)then
            tdeb=tmin-a*drnunf()
            nt=ifix(sngl((tmax-tdeb)/a))+1
            do it=1,nt
              t(it)=dfloat(2*idnint(drnunf())-1)
            enddo
c
            do ix=1,nx
              x=xmin+(ix-1)*echx
              t0=x*cdx(id)+y*cdy(id)+z*cdz(id)
              dt=dmod(t0-tdeb,a)
              it= dint((t0-tdeb-dt)/a+1.5d0)
              buf(ix)=buf(ix)+t(it)*(2.d0*dt/a-1.d0)
            enddo
          endif
c
c  simulation in the exponentiel case
c
          if (struct.eq.2)then
            b=2.d0*a
            t(1)=tmin+b*dlog(drnunf())
            t(2)=tmin-b*dlog(drnunf())
            nt=2
            do while (t(nt).le.tmax)
              nt=nt+1
              t(nt)=t(nt-1)-b*dlog(drnunf())
            enddo
c
            do ix=1,nx
              x=xmin+(ix-1)*echx
              t0=x*cdx(id)+y*cdy(id)+z*cdz(id)
              itp=1
```

```
          itn=nt
          do while (itn-itp.gt.1)
            it=(itp+itn)/2
            if (t0.ge.t(it))then
              itp=it
            else
              itn=it
            endif
          enddo
          if(2.d0*t0.gt.t(itp+1)+t(itp))then
            buf(ix)=buf(ix)-1.d0
          else
            buf(ix)=buf(ix)+1.d0
          endif
        enddo
      endif
c
c  simulation in the stable case
c
      if (struct.eq.3)then
        call drnnoa(1,b)
        b=2.d0*a*b*b
        call drngam(1,0.5d0,c)
        call drnnoa(1,ux)
        call drnnoa(1,uy)
        call drnnoa(1,uz)
        r=dsqrt((ux*ux+uy*uy+uz*uz)/2.d0/c)/b
        phi=2.d0*pi*drnunf()
c
        do ix=1,nx
          x=xmin+(ix-1)*echx
          t0=x*cdx(id)+y*cdy(id)+z*cdz(id)
          buf(ix)=buf(ix)+dcos(r*t0+phi)
        enddo
      endif
c
c  simulation in the hyperbolic case
c
      if (struct.eq.4)then
        u=1.
        do while (u.gt.drnunf())
          u=dsqrt(drnunf())
        enddo
        t(1)=tmin-a*drnunf()*(1.d0/u-1.d0)
        tv(1)=dfloat(2*idnint(drnunf())-1)
        nt=1
        do while (t(nt).le.tmax)
          nt=nt+1
          t(nt)=t(nt-1)+a*(1.d0/drnunf()**0.3333-1.d0)
          tv(nt)=dfloat(2*idnint(drnunf())-1)
        enddo
c
        do ix=1,nx
          x=xmin+(ix-1)*echx
          t0=x*cdx(id)+y*cdy(id)+z*cdz(id)
          itp=1
          itn=nt
          do while (itn-itp.gt.1)
            it=(itp+itn)/2
            if (t0.ge.t(it))then
              itp=it
            else
              itn=it
            endif
          enddo
          buf(ix)=buf(ix)+tv(itp)
        enddo
      endif
c
c  simulation in the gaussian case
c
      if (struct.eq.5)then
        call drnnoa(1,ux)
        call drnnoa(1,uy)
        call drnnoa(1,uz)
        r=dsqrt(2.d0*(ux*ux+uy*uy+uz*uz))/a
        phi=2.d0*pi*drnunf()
c
```

```
            do ix=1,nx
              x=xmin+(ix-1)*echx
              t0=x*cdx(id)+y*cdy(id)+z*cdz(id)
              buf(ix)=buf(ix)+dcos(r*t0+phi)
            enddo
          endif
c
c  simulation in the cardinal sine case
c
          if (struct.eq.6)then
            r=dfloat(2*idnint(drnunf())-1)/a
            phi=2.d0*pi*drnunf()
c
            do ix=1,nx
              x=xmin+(ix-1)*echx
              t0=x*cdx(id)+y*cdy(id)+z*cdz(id)
              buf(ix)=buf(ix)+dcos(r*t0+phi)
            enddo
          endif
c
c  end of loop on the directions
c
        enddo
c
c  store the line
c
        do ix=1,nx
          buf(ix)=buf(ix)*sigma
          vmax=max(vmax,buf(ix))
          vmin=min(vmin,buf(ix))
        enddo
        write (40,2100)(buf(ix),ix=1,nx)
c
c  end of loop on the y and z coordinates
c
        write (6,3100)cr,iz,iy
      enddo
      enddo
c
c----------------------------------------------------------
c
c  end of program
c
c
      write (6,3200)cr
      write (6,*)'end of program'
c
      write (6,*)'minimum',vmin
      write (6,*)'maximum',vmax
c
c  this is the end...
c
      stop
      end

c==========================================================
c==========================================================

      subroutine vdc(nv,nvmax,ux,uy,uz)
c
c
c  Generate a family of unit vectors (van der Corput sequence)
c
c  Input arguments
c  - nv      number of vectors to be generated
c  - nvmax   maximum number of vectors allowed
c
c  Output arguments
c  - ux      array containing the x coordinates of the vectors
c  - uy      array containing the y coordinates of the vectors
c  - uz      array containing the z coordinates of the vectors
c
c
      implicit real*8 (a-h,o-z)
c
      dimension ux(nvmax),uy(nvmax),uz(nvmax)
c
c     common iseed
```

```
c
c
c-----------------------------------------------------
c
c  generate vectors
c
      if (nv.gt.nvmax)stop   'too many vectors to be generated'
c
c  loop on the vectors
c
      do iv=1,nv
c
c  binary decomposition of iv
c
        n=iv
        x2=0.d0
        d=2.d0
        do while (n.ne.0)
          m=mod(n,2)
          x2=x2+dfloat(m)/d
          d=d*2.d0
          n=n/2
        enddo
c
c  ternary decomposition of iv
c
        n=iv
        x3=0.d0
        d=3.d0
        do while (n.ne.0)
          m=mod(n,3)
          x3=x3+dfloat(m)/d
          d=d*3.d0
          n=n/3
        enddo
c
c  coordinates of the vector
c
        ux(iv)=dcos(6.283185d0*x2)*dsqrt(1.d0-x3*x3)
        uy(iv)=dsin(6.283185d0*x2)*dsqrt(1.d0-x3*x3)
        uz(iv)=x3
c
c  end of loop on the vectors
c
      enddo
c
c-----------------------------------------------------
c
c  random rotation of the vectors
c


c
c  generate the rotation axis
c
      call drnnoa (1,ax)
      call drnnoa (1,ay)
      call drnnoa (1,az)
      r=dsqrt(ax*ax+ay*ay+az*az)
      ax=ax/r
      ay=ay/r
      az=az/r
c
c  generate the rotation angle
c
      theta=6.283185d0*drnunf()
c
c  rotation of the vectors
c
      do i=1,nv
        call rotation (ux(i),uy(i),uz(i),ax,ay,az,theta,
     &                 ux(i),uy(i),uz(i))
      enddo
c
c-----------------------------------------------------
c
c  end of subroutine
c
      return
```

```
      end
c=====================================================
c=====================================================

      subroutine rotation (x0,y0,z0,ax,ay,az,theta,x1,y1,z1)
c
c
c  Rotation of a point in three-dimensional space
c
c  Input arguments
c  - x0,y0,z0   the coordinates of the point before the rotation
c  - ax,ay,az   direction of the rotation axis
c  - theta      rotation angle
c
c  Output arguments
c  - x1,y1,z1   the coordinates of the point after the rotation
c
      implicit real*8 (a-h,o-z)
c
c----------------------------------------------------
c


c
c  projection of the point onto the rotation axis
c  (carried by unit vector a)
c
      r=x0*ax+y0*ay+z0*az
      px=r*ax
      py=r*ay
      pz=r*az
c
c  stop if the point belongs to the rotation axis
c
      if ((x0.eq.px).and.(y0.eq.py).and.(z0.eq.pz))then
        x1=x0
        y1=y0
        z1=z0
        return
      endif
c
c  produce a unit vector b which is orthogonal
c  to the rotation axis
c
      r=dsqrt((x0-px)*(x0-px)+(y0-py)*(y0-py)+(z0-pz)*(z0-pz))
      bx=(x0-px)/r
      by=(y0-py)/r
      bz=(z0-pz)/r
c
c  produce a unit vector c which is orthogonal to a and b
c
      cx=ay*bz-az*by
      cy=az*bx-ax*bz
      cz=ax*by-ay*bx
c
c  rotation in the plane spanned by b and c
c
      ct=dcos(theta)
      st=dsin(theta)
c
      x1=px+r*ct*bx+r*st*cx
      y1=py+r*ct*by+r*st*cy
      z1=pz+r*ct*bz+r*st*cz
c
c  end of subroutine
c
      return
      end
```

DISCUSSION FOLLOWING SESSION NO 5

Chairman: Peter Dowd

Papers: de Fouquet; Lantuéjoul

After the end of both papers, the chairman called for questions.

Omre — Efficiency is a very important aspect in simulation. Could you tell us something about how fast your algorithms are ?

Lantuéjoul — To simulate an exponential variogram on a 400 × 400 grid (i.e. 160 000 pixels) it takes less than 15 minutes on a SPARC 2 station.

Omre — That is a lot of time for a gaussian process.

Lantuéjoul — Probably this is not a very good example.

Omre — I think your approach is interesting, from an academic point of view but to be quite honest I do not think the idea will work in practice if efficiency is the criterion. It is so simple to simulate multigaussian processes (e.g. using sequential methods), that I do not think that using these processes that converge in the limit will be practicable, because sequential methods go straight to the result.

Lantuéjoul — When we say that we want to simulate a multigaussian process, what does this mean ? By multigaussian I mean that the model is fully specified.

Omre — And you have to do this to simulate.

Lantuéjoul — In practice we never know the spatial distribution of the random function. At most we know the 4th order moments. So where should we stop ? Where is the cutoff ? For instance for a moment of order 4, it turns out that about 60 lines should be sufficient. Coming back to the 15 minutes of computer time required for a spherical, I was considering 400 directions.

Omre — I would never simulate unless I had made an explicit assumption (perhaps a subjective one) about the distribution. So I agree with you about knowing it exactly. If you do not know it, you have to assume it.

Armstrong — The computer time to run 2 simulations of 120 000 points is quite quick (less than 5 minutes). Time is not really a problem.

M. Armstrong and P. A. Dowd (eds.), Geostatistical Simulations, 178–184.

Galli — The examples presented were all in 2D, whereas I am interested in 3D. What difference in time would you expect for a 3D grid with the same number of points ?

Lantuéjoul — What is important, is the average diameter of your simulation field; I mean the projection of the simulation field over the lines. Except for very elongated fields, things are not very different going from 2D to 3D.

Ravenscroft — You said you were limiting yourself to the isotropic case. Are we still in the situation where if we have a double structured variogram or zonal anisotropy, we have to do separate simulations for each structure ?

Lantuéjoul — Firstly for the anisotropy, I have only considered the isotropic case because geometric anisotropy simply corresponds to a dilation of space and you come straight back to simulating an isotropic structure. Zonal anisotropy amounts to carrying out a simulation in a subspace. So if I want to consider anistropy, it should be more general types of anisotropy.

Now for your second question about the sum of 2 structures, this is a question of how the program has been written. For the user it is better to have the 2 simulations within the program (rather than running it twice).

Renard — In the earlier ways of implementing turning bands, one had to discretize the bands. Can you tell us about this ?

Lantuéjoul — I prefer avoiding any discretization unless it is completely unavoidable. With discretization, there is always a risk of changing the simulation model, and therefore I always prefer to work in a continuous space. All of the spectral and turning band methods presented in the paper have been defined in a continuous space. In fact it is simpler to work in a continuous space.

Gomez — What worries me – more than the speed of the computation and convergence of the covariance – are the lines, these artefacts that I showed yesterday. When you compute the 3D variogram map, it often turns out very badly, even if the average variogram is good. You can see quite different variograms along the lines.

Lantuéjoul — You said that your simulation was carried out with only 16 lines.

Gomez — In your results, we could see the lines in the maps let alone the variogram map.

Lantuéjoul — The point is that the variogram is just a second order moment. So this is negligible compared to the whole structure of the distribution.

Armstrong Wasn't the simulation that you showed to us a particularly bad one ? You said that the method used was suboptimal ? The one you used to get the last few slides.

Lantuéjoul It was an exponential simulation using the spectral method.

Armstrong Would you recommend that we use it ?

Lantuéjoul No, not at all. I prefer to use the migration method. It is very easy to implement.

Dowd As the spirit of the meeting was meant to be a workshop, I would like to ask Jaime Gomez about the sequential methods. Could you answer two points that Christian Lantuéjoul raised in the written version of his paper about sequential methods ? These are (1) numerical instabilities when the covariance is very smooth at the origin and (2) the fact that it cannot be applied directly when there are more than 500 points. This is presumably because of the necessity to use a moving neighbourhood or to select points within the neighbourhood, and the problems of inverting large matrices. Perhaps Christian Lantuéjoul could expand on this before Jaime Gomez replies.

Omre We have been working on this.

Gomez We were already discussing this yesterday. First of all, the limitation to 500 nodes is only if you cannot get the conditional distribution exactly. You can always approximate it by using the closest points within a search neighbourhood. That works very well for covariances like the exponential or the gaussian, but not as well for the spherical. For that, you have to take more points and a larger neighbourhood. But it still works fine.

Now for the numerical instabilities, if the separation between points within the search radius is very small compared to the range of the gaussian covariance, you can have numerical instabilities because all the covariances in the kriging matrix are almost equal to 1.0, up to machine precision. One way to fix that is by adding a little nugget effect. Of course, you would not reproduce the multivariate gaussian model exactly then. However it does work.

Lantuéjoul Just a short question. Have you tried to quantify the effect of the approximation when you do not take all the points into account but only part of them ?

Omre Have you read the Troia paper ? It is given there. *(Ed: He is referring to "Simulation of random functions on large lattices" by Omre Solna & Tjelmland)* In 1D you can compute a lot of things exactly because of the

Markov properties. In 2D, the Markov properties are harder to treat. You have to rely on approximations. We have been working on the exponential, to test this. It is reported in the Troia paper. It works surprisingly well. We have been trying to build from 1D where things are well known, up to 2D and 3D. I will just say that as long as you have exponential variograms you are fairly safe.
And it is very fast. I would be prepared to do a test run with you on any machine you like, and for any grid size.

Dowd It's a simulation Grand Prix.

Gotway From what we have been hearing, there seem to be only two criteria for defining a good simulation : speed and reproducing the variogram. I would contend that they should not be the only ways for judging algorithms. I have seen simulations satisfying both criteria, but after the transfer function is calculated the shapes of the uncertainty distributions are different. So you have to look at what you intend to do with the simulations. You should really judge the simulation algorithm by looking at the end result.

Lantuéjoul This is why we are considering not just the variogram but the whole spatial distribution of the function.

Gotway Coming back to the striping problem with the turning bands, in this study we did not see any realizations that did have banding. In other studies when banding was present, it did not seem to affect the transfer functions. The fields may not be visually appealing but in terms of what you are going to do with them, it does not seem to be affecting the nature of the uncertainty distributions.

Lantuéjoul I showed you some slides of simulations with exponential, spherical, gaussian, stable models. Did you see any banding there ? Any lines in them ? So maybe it is possible to avoid these.

Gotway It is possible to avoid the lines. But even when they are there, what effect do they have on the **inferences** that are being made based on the realisations ? That is the key question.

de Fouquet Sequential methods generate simulations in a step by step way. In your model you are using something that does not depend on the way you make the simulation but in practice you simulate a point, then a second conditional to the first and so on.. The order of taking the points could influence the results. Have you checked for differences in the spatial distribution due to this lack of symmetry ?

Gomez — In principle, it would not matter if you could compute the conditional disribution exactly at all nodes. But you have to limit the number of points used to approximate the cdf. The sequential path creates artefacts; it is too artificial. However you do not see any artefacts using a random path, as I showed yesterday. It not merely reproduces the covariance; the bivariate distribution is gaussian. To my knowledge, it is multigaussian. So coming back to Carol Gotway's comments, I feel that the width of uncertainty that she got is not due to the sequential simulations; instead it results from the multigaussian with all the consequences about maximum entropy and low connectivity of extreme values that such a model implies. It is not due to the algorithm...

Gotway — That is not true. I generated an exhaustive multigaussian data set and applied two gaussian based simulation methods and the shapes of the resulting uncertainty distributions changed. So I know that I am dealing with a multigaussian data set.

Gomez — Maybe you did not do enough realisations.

Gotway — I did 100 of them. Maybe that is not enough but practically speaking it is a lot.

Dowd — Carol is making a point about the use of the simulations. In mining applications where I have seen simulations tried and fail, it has been for exactly this reason. The purpose of the simulation has not been taken into account, and has not been tested. So a large number of very valid applications of financial analysis where the idea was to apply simulations to an orebody drilled out with very sparse data and a whole range of financial scenarios ; worst, best, middle. But if the simulation did not produce the total range of possible pictures of that deposit, so they totally fail. People do not publish such studies but people in this room must have seen them. They have seen millions of pounds resting on a decision made on these sorts of simulations.

The chairman then threw the discussion open to questions and comments on both papers.

Ravenscroft — My point is actually general but perhaps Chantal de Fouquet can best answer it. In the classical approach to turning bands (and in the way all multigaussian based simulations are proposed), there is normally a step to ensure that the simulated gaussian values are N(0,1). It can be applied line by line or to the whole simulation. In my experience, this could lead to the decrease in the variability in the response distribution because you are not taking all the generated randomness ; you are trying to squeeze back into a N(0,1) distribution. Hence my question. Are we still doing that ? Is it correct ? Because if you are simulating a very small field relative to the variogram...

de Fouquet If the field is large enough the simulation should respect the statistical properties of the model. But for a small field you are quite right. You have to calculate the dispersion variance for the domain to be simulated. If the domain is big enough, you have the theoretical dispersion variance. If not, it would not be correct to standardize.

Ravenscroft Let us assume that the dispersion variance is 0.8 ? Are we still right in bending all our realisations to match exactly the same gaussian distribution ? Are we not taking out some of the variability between simulations before we start ? We are removing some of the randomness introduced at the outset.

Gomez If you believe that the sample statistics are the absolute truth and that each realization should reproduce them you should restandardize, but it is more realistic to assume that the sample could be biased one way or another, in which case it is better to let the realization statistics fluctuate.

Omre You are simulating a gaussian (0,1) process. So if you take your realisation and square all your numbers, that is, subtracting zero, it should be 1.
You must not subtract the average of your values.

Ravenscroft My question is philosophical. Should we be standardizing ?

de Fouquet If the domain is short compared to the range, then you should not. One can carry out the following exercise: carry out a simulation on a large enough area so that the dispersion does equal the variance, then reduce the size. Clearly the dispersion of the simulated values will be less than the model variance.

Allard A question for Henning Omre. I have read the Troia paper, but perhaps I have missed something. When you tested your sequential simulations, your test was on the variogram.

Omre But we have done other tests, on all sorts of aspects.

Allard Do you have any ideas as to how far the simulations deviate from the gaussian assumption ?

Omre I am always working in a gaussian world ; drawing gaussian random variables, making linear combinations of them. That gives me some guarantees...

Allard If you approximate the conditional law, you find you lose some properties.

Omre I may lose some spatial distribution but not multigaussianity.

Allard If you want to make a multigaussian simulation to calculate some complex nonlinear function, you want to ensure that your simulation is the right multigaussian and not just an approximation. If it is an approximation, you need to know how far it is from the real one.

Omre The hard part is to get the values high up in the tails. By controlling only the 4th moment and by invoking the theorem it is a much harder task than by having gaussian variables all the time. So I think it is theoretically correct, but with sequential methods, you stay in the gaussian domain all the time.

Allard But we still have to work out why two apparently gaussian generating algorithms produce such different responses for the transfer functions.

Omre The objective is clear– having a multigaussian realization. That is the model. We need to know how much weight to put on different deviations.

Lantuéjoul Suppose you want to use a 10 point neighbourhood for a sequential simulation. Is it possible to find a suitable expression for the distribution of a linear combination of 11 variables ? You are dealing with a development of just 10 points.

Omre We have got various approximations. We know exactly what happens in 1D. There are many approximations in 2D. You have to consider the configuration rules and not just the number of points. I admit. So an empirical evaluation would probably be the best. But what criterion should we use?

There was one thing in Chantal's talk that concerned me. You can do prediction, and find the best one under the risk of square error loss but when you are simulating, you have to give yourself the complete probability model like you do with the gaussian. It is not sufficient to test for binormality. If you cannot test it, you have to make the assumption that you know it exactly. The model has to be stated clearly.

de Fouquet Earlier I talked about the conditioning from the experimental data. Using the experimental data if they are on a regular grid, you can try to calculate higher order moments, but if they are irregular, you try to check the bivariate distribution experimentally but you can rarely do more.

MODELLING THE KARSTIC MEDIUM: A GEOSTATISTICAL APPROACH

O. JAQUET
Colenco Power Consulting Ltd.
CH-5405 Baden

P.Y. JEANNIN
Centre d'Hydrogéologie
de l'Université de Neuchâtel
CH-1200 Neuchâtel

Karstic medium is an heterogeneous medium, characterized by the presence of a network of conduits that strongly influences groundwater flow. The elaboration of a karst model is necessary for the understanding of karstic groundwater flow. Based on the density of conduits, a geostatistical approach is proposed for the modelling of karstic medium. This approach satisfies the constraints required by hydrogeological applications, but some difficulties related to the model connectivity need further evaluation.

INTRODUCTION

From a hydrogeological point of view, the karstic medium, also called karst, is a heterogeneous medium composed of low-permeability rocks (matrix), with water-conducting fractures and highly permeable conduits (karstic networks). In the saturated zone of karstic aquifers, the network of conduits drains the fissures and the rock matrix toward a discharge area. The geometry of the highly permeable drainage network considerably influences the hydraulic behaviour and the transport of pollutants within the entire karstic aquifer.

A full understanding of groundwater flow cannot be gained by simply applying an effective permeability of the karstic medium (Kiraly, 1976), but rather requires the elaboration of a model of the karst. Due to the complexity of the network geometry, and to the generally small amount of available geological information, a probabilistic approach was chosen for modelling the karstic medium.

An empirical geostatistical-simulation method is proposed that allows the generation of images of the spatial layout of conduits within the karstic medium. The method is applied to the Hölloch site in Switzerland. This site is presently being investigated within the framework of the European project COST 65, which concerns the protection and the management of groundwater in karstic media. This

M. Armstrong and P. A. Dowd (eds.), Geostatistical Simulations, 185–195.

note also defines the scope of a proposed research effort that is aimed at modelling the karstic medium using geostatistical methods.

THE HÖLLOCH KARSTIC NETWORK

The Hölloch is the largest known cave network in western Europe. With 160 kilometres of tunnels known today, the Hölloch is the third longest network in the world. This karstic network forms a labyrinth of conduits that develop along a plane dipping toward the NNW (figure 1). The oldest conduits are at the highest altitude; they were formed in the saturated zone, at a time when the level of the neighbouring valleys was higher than today. With time and the deepening of the valleys, the altitude of the sources decreased and new conduits were formed at lower levels.

As they moved downward, the conduits followed an inclined plane that was favourable to their formation, and therefore migrated toward the NNW. Presently, groundwater flow is from east to west in saturated conduits within a band, which has a width of half a kilometre under low-water conditions, in the NNW part of the karstic network. During flooding periods, groundwater level rise in the conduits of the lower part of the karstic network.

In a first step, the analysis of the network was limited to an area that includes conduits that were formed under conditions similar to those of the conduits of the saturated zone, where groundwater occurs today. However, the analysed network includes the conduits that formed during the development of at least three successive flow systems in a vertical section of 350 metres.

Generally, the available data concerning the geometry of karstic networks are either minimal or lacking completely. The Hölloch site is well documented and therefore is particularly well suited to help define and test models of karstic media.

DATA OF THE HÖLLOCH

Data concerning the Hölloch karstic network were collected by speleologists in conduits of at least 0.3 metre in diameter. This group of data forms a sub-group of the total karstic drainage network, which is the group of existing conduits for any size. In a first stage, the small conduits are not considered. However these conduits (with a diameter less than 0.3 metre) contribute substantially to flow and they will therefore be included in the karst model at a later stage.

The explored conduits are defined by straight segments that were measured with a tape; directions were measured with a compass, and slopes were measured with a clinometer. The uncertainty attached to the data is approximately 1 to 2 % of the measured length (Jeannin, 1992). From the group of identified conduits, the mapping of the karstic network was performed (figure 1).

The diameter of the conduits was also measured. In a first stage, this variable, which is required for modelling of groundwater flow, is not taken into consideration for the characterisation of the karstic network.

For technical reasons, the Hölloch data are not yet available in digital form, and we therefore worked directly on the map of the karstic network. This approach implies considering an area of the network where the density of the conduits is considered approximately homogeneous in the vertical dimension (figure 1: network with bold lines). The map is a projection on a horizontal plane of a network that is developed in the three-dimensional space. The apparent density of the conduits measured on the map is therefore overestimated with respect to the true density of the conduits in space.

DENSITY OF CONDUITS

The spatial variability of the geometry of the karstic network was studied with the help of the following regionalized variable: the sum of the conduit lengths within a square block of fixed size. This two-dimensional variable describes the density of conduits for a given support:

$$d_v(x) = \sum_i lc_i/v \tag{1}$$

$d_v(x)$: density of conduits $[m/m^2]$
lc_i: length of conduit i belonging to block v.

The structural characteristics of the variable depend directly on the choice of the size of the measurement support. The selection was carried out in order to allow the acquisition of a large enough number of data while avoiding a too large proportion of zero values. The selected dimensions of support for the Hölloch karstic network are 200 by 200 metres. The influence of the size of the support on the spatial behaviour of the variable will be examined once the data become available in digital form.

The map of the karstic network was discretized as conduits of 50 metres in length, in order to facilitate the sampling of density of the conduits. Because the length of the conduits is fixed, according to equation (1), the number of conduits per block is the sampled variable:

$$ncb(x) = (d_v(x) \cdot v)/lf \tag{2}$$

$ncb(x)$: number of conduits per block v [-]
lf: conduit length.

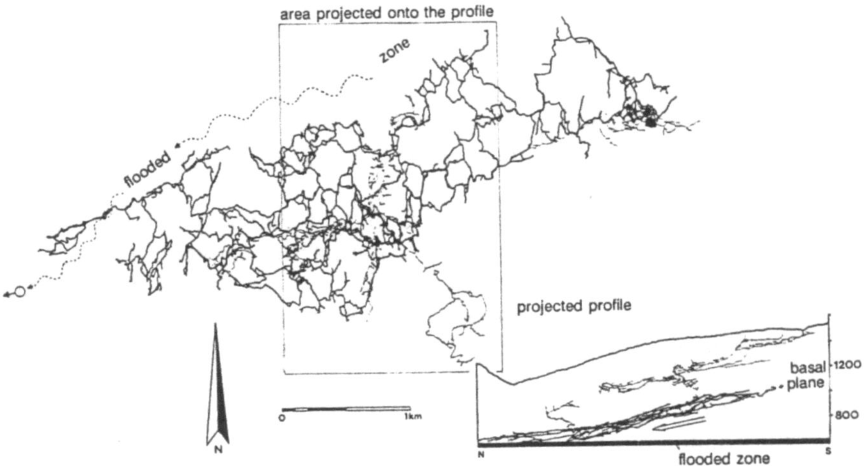

Figure 1 Map of the Hölloch karst (Rouiller & Auf der Maur, 1986)

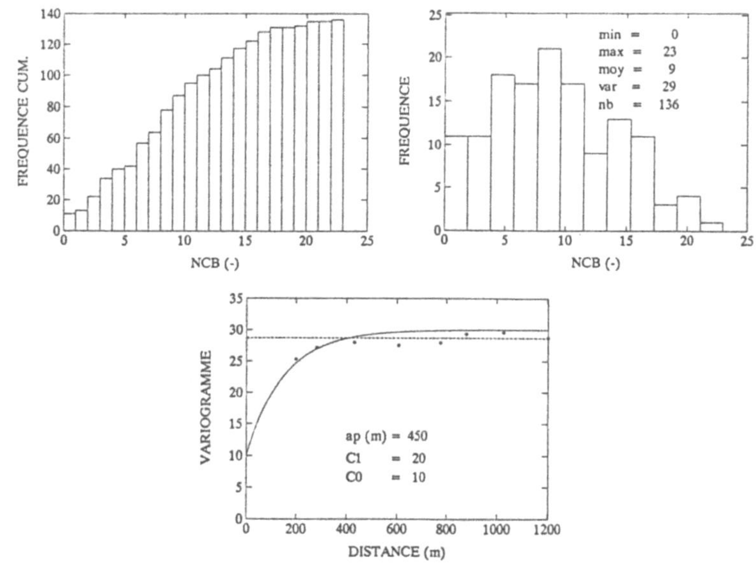

Figure 2 Statistics of the number of conduits per block

The number of conduits per block was obtained by moving the selected support along a regular grid placed on the map of the karst, and manually counting the number of 50-metres conduits that are intersected by the support at each node.

Histogram

The histogram of the number of conduits per block (strictly, the number of 50-metres conduits contained within the support v) displays an asymmetrical tendency that indicates a larger proportion of low values than high values for the studied variable (figure 2).

Due to the provisional status of the data, a transformation of the variable (lognormal, anamorphosis), which allows the conditioning of the simulation results to the histogram of the data, was not performed. Once the data become available, the variables characterizing the conduits (length, direction, slope and diameter) will be analysed in detail.

Variogram

The behaviour of the experimental variogram (isotropic) of the number of conduits per block shows that this variable can be considered as stationary (figure 2). The experimental variogram is weakly structured. The reasons for this effect are the following:

- the size of the selected support (200*200 m) is too large, and
- due to the manipulations to which they were subjected before variogram analysis, the data probably were destructured.

To proceed in the exploratory study while waiting for data of better quality, the experimental variogram was fitted by an isotropic exponential model with a correlation scale of 450 metres. The chosen nugget effect corresponds approximately to the order of magnitude of the measurement error of the conduit lengths.

GEOSTATISTICAL APPROACH

Model constraints

The construction of a karstic medium model that is adapted to hydrogeological applications was undertaken with various constraints:

- the model respects the available geological information;

- the two-dimensional geometry of the model is compatible with numerical methods for flow modelling;

- the geometry of the model is hybrid (two types of elements), consisting of a network of geometrical elements (conduits) that are contained within a rock matrix; and

- it should be possible to extend the two-dimensional geometry of the model (chosen for parsimony), to three dimensions.

The combination of these imposed constraints allows the definition of the selected karst model:

- a two-dimensional model,
- fitted to the geological data and,
- with a geometry of finite-element type.

KARST MODEL

The model allows the simulation of karstic network images that preserve the observed statistical characteristics (mean, variance and correlation scale). The simulation of the karstic network was carried out in two stages.

a. Number of conduits

The simulation of the number of conduits per block was carried out by the turning bands method (Chilès, 1977) with the help of the code TUBA (Zimmerman and Wilson, 1990). The number of conduits per block (stationary variable) was simulated at centres of blocks that form a regular grid of similar extension to the real network. The size of blocks equals that of the variable support. For each block of the karstic image, the number of conduits with a length of 100 metres (chosen for parsimony) was obtained after a transformation of the simulated values into integers varying from zero to twelve.

b. Geometry

The conduits must be placed within the network blocks. After simulating the values of the variable, conduits were randomly (with an equal probability) allocated within each block of the network (figure 3).

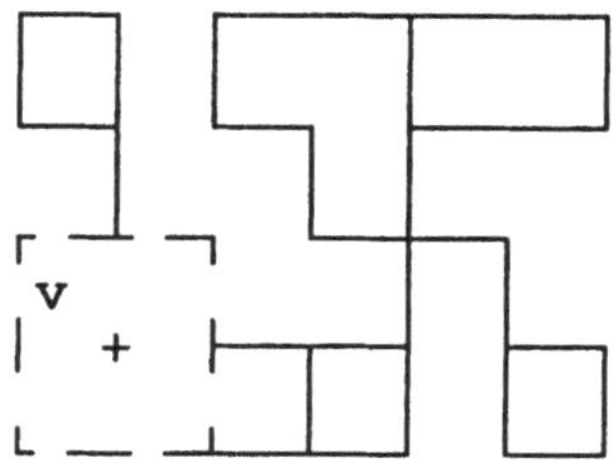

Figure 3: conduits of the network (lines)

The obtained image of the karst consists of two groups:

- the karstic network, which assembles the conduits, and
- the rock matrix within the blocks that surround the karstic network.

To limit the number of geometrical elements (conduits and blocks) that make up the model, only two orientations (N-S and E-W) and twelve possible positions for the conduits within each block are allowed. This choice aims at avoiding large computer times when modelling groundwater flow using the finite-element numerical method.

The obtained model of the karstic medium resembles a boolean type of model with a random regionalized density (punctual process of Cox), whose primary grains are formed by the conduits of the network (Lantuéjoul, 1993). This model differs from the boolean one by the constraints imposed on the orientations and the positions of the conduits within the medium. The consequences of these constraints on the theoretical formulation of this random set model have not yet been established.

Karstic images

With the help of the model, four images (or realisations) of the Hölloch karstic network - rectangular zone of 6 by 3 kilometres - were generated (figure 4). When the number of conduits per block was simulated, only the structured part of the variogram was taken into account. The nugget effect, interpreted as measurement errors, was not simulated, the superposition of a nugget component to the structured part only adds noise to the resulting images.

The four realisations differ from the map of the network (figure 1) by a large apparent homogeneity of the density of the simulated conduits in space. This effect is due to the smoothing manipulations to which the data were subjected during their sampling.

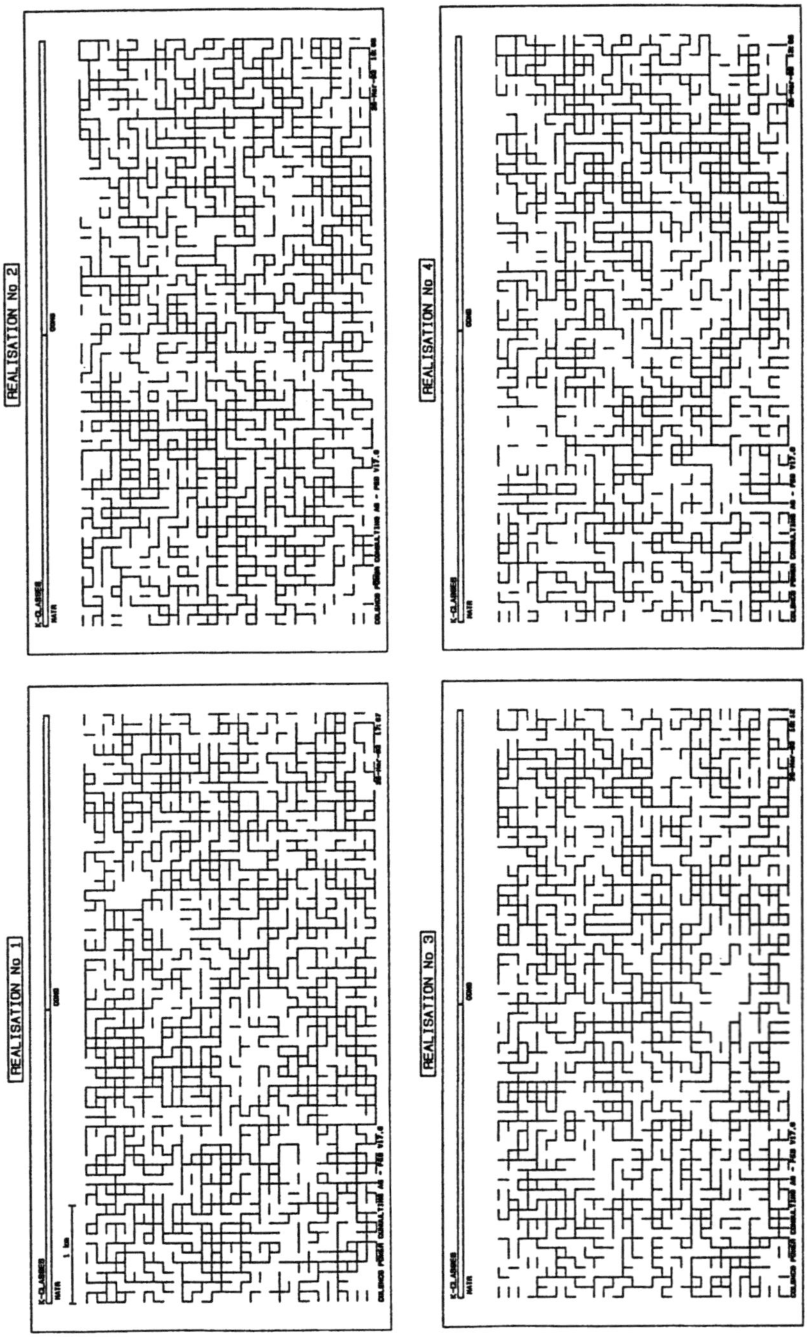

Figure 4 Geostatistical simulations of the Hölloch karst

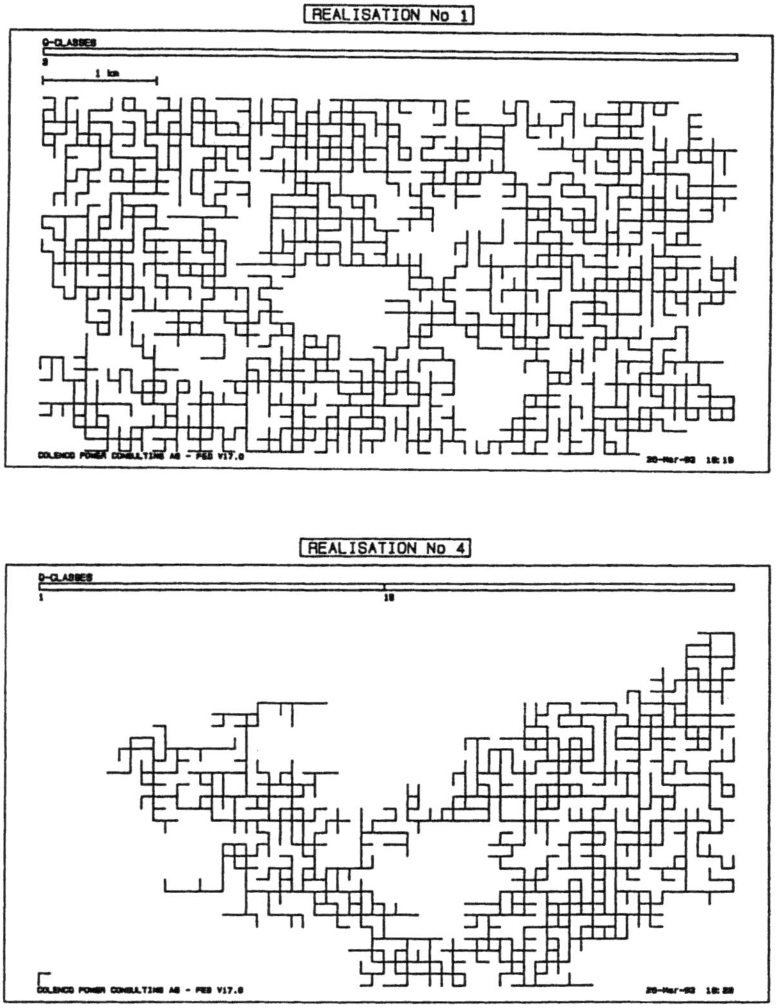

Figure 5 Network extent and connectivity

The network images also display groups of isolated conduits, not connected to the network, which in principle do not exist in reality. The area of the network (forming the group of connected conduits) is highly variable from one realisation to an another (figure 5).

The connectivity of the realisations is influenced by the density of karstic conduits and by the scale of the regionalized phenomenon (Allard, 1992). These characteristics alone are not sufficient for the determination of the connectivity, which is a consequence of the distribution law of the selected random set model. A random set model adapted to karst should include a connectivity parameter, estimated from the data. In parallel, the connectivity of the images could be improved by incorporating the spatial correlation of the conduits' orientation in the model.

The proposed simulation method is not conditional. With respect to the karstic medium, the simulations should be conditional. The amount of information related to the geometry of the conduits generally is sparse, but the flow directions are relatively well known and so are the discharge zones (springs).

Based on the network data and on the geological knowledge of the area, the hydrogeologist should first outline the karstic network. Using this information, a conditional simulation method with external drift could be performed. The karstic network would be simulated conditionally to the geometrical data and to the outline of the network, used as an external drift. This new approach is currently being studied.

PRELIMINARY CONCLUSIONS AND PERSPECTIVES

The proposed geostatistical simulation method is a preliminary approach at modelling karstic medium and several further developments are still to be undertaken.

This method has the advantage of satisfying the constraints required by hydrogeological applications, and, furthermore, its extension to three-dimensional space is feasible. The analysed regionalized variable, the density of conduits, is well-suited for the characterisation of the spatial variability of the density of the karstic network.

However, this approach does not directly address the issue of observed network connectivity. Further developments are necessary: the conditionalisation of the obtained realisations will certainly lead to an improvement of the connectivity of the model. A karstic model that intrinsically includes the notion of connectivity allows a better description of the observed reality.

While waiting to apply this method on the raw data, other approaches for modelling karst are actively investigated.

ACKNOWLEDGMENTS

We thank the speleologists of the AGH (Arbeitsgemeinschaft Höllochforschung), who map the caves and carefully record the documents that were made available to us. This note was written within the context of the project entitled "Simulation of coupled flows in fissured and karstic media (subside No 20-230305.90) and which is funded by of the Swiss National Fund for Scientific Research.

REFERENCES

Allard, D. (1992) "On the connectivity of two random set models: the truncated gaussian and the boolean", proceedings of the 4th international geostatistics conference, Troia, Portugal.

Chilès, J.P. (1977) "Géostatistique des phénomènes non stationnaires (dans le plan)", Université de Nancy, thèse, 1-152p.

Jeannin, P.Y. (1992) "Géométrie des réseaux de drainage karstique: approche structurale, statistique et fractale", annales scientifiques de l'Université de Besançon, mémoire hors série, no 11, 1-8p.

Kiraly, L. (1976) "Remarques sur l'hydrogramme des sources karstiques simulé par modèles mathématiques", bulletin du centre d'hydrogéologie, Neuchâtel, vol. no 1, 37-60p.

Lantuéjoul, C. (1993) "Ensembles aléatoires", centre de géostatistique, Fontainebleau, France, 1-33p.

Rouiller, P. & Auf der Maur (1986) "Übersichtplan des Höllochs 1:25000", Stalactite No 36, 28-29p.

Zimmerman, D.A. and Wilson, J.L. (1990) "Description of and user's manual for TUBA: a computer code for generating two-dimensional random fields via the turning bands method", Seasoft, Albuquerque, 1-155 p.

Simulating a Geological Lithofacies with Respect to Connectivity Information Using the Truncated Gaussian Model

Denis ALLARD
Centre de Géostatistique — Ecole des Mines de Paris
35 rue Saint-Honoré, 77305 Fontainebleau, France

ABSTRACT : One of the main parameters in oil and gas reservoir exploitation is the connectivity of the permeable phase. Generally, conditional simulations of the geological lithofacies do not take into account the available information about connectivity that can be provided by well-test or geological interpretation for instance. Using the truncated gaussian model, an iterative simulation algorithm is proposed which respects connectivity constraints such as :

a) x and y *must* be connected.
b) x and y *must not* be connected.

Practical cases will illustrate the consistency between the model of the gaussian random function and the constraints of the conditional simulations.

1 Introduction

To give numerical models of heterogeneous oil and gas reservoirs that can both reproduce the heterogeneity and honor the maximum of data and available information, is one of the major aims of petroleum geostatistics. For this purpose, several stochastics models have been proposed during the past decade, including truncated gaussian model [11] (TGM), the sequential indicator simulation (SIS) [1], and the boolean model. The last one does not allow correct conditional simulation yet, when the SIS algorithm honor "hard" and "soft" data, i.e. exact and imprecise values. With the TGM model, one can today perform simulations conditionally to "hard" and/or "soft" data, as well as noisy [6] or regularized ones [5]. With a multivariate generalization, a cosimulation can be used to take into account the information given by another extensively sampled variable. An example is the cosimulation of porosity using the seismic absolute acoustic impedance [12]. Although all these improvements integrate increasing amounts of information, they do not take into account one of the essential parameters in flow simulation : the connectivity of the permeable lithotype.

M. Armstrong and P. A. Dowd (eds.), Geostatistical Simulations, 197–211.

An attempt has been made in [10], but only a local, particular geometrical aspect was considered. More recently, a simulated annealing type algorithm which respects well-test curves has been proposed [3].

In this paper, we propose an algorithm to simulate the TGM under connectivity constraints such as "x is connected to y", and/or "x is not connected to y". This algorithm is derived from the Gibbs-sampler, introduced in 1984 by Geman and Geman [7].

The connectivity of the random set depends very much on the model and on the grid on which the simulation is performed. Therefore, after having briefly recalled some definitions concerning the TGM and connectivity on a regular simple square grid, we will illustrate the influence of the model of the random set and of the range parameter on some connectivity characteristics in section 3. Then, we will present formally the algorithm of conditional simulation (section 4). Before concluding, we will illustrate it by practical cases and get on to the problem of the adequation between the prior model and the constraints.

2 Definitions

On the square grid $G = \{1, \ldots, n\} \times \{1, \ldots, n\}$, each site $x = (i,j)_{1<i,j<n}$ has 4 nearest-neighbours : $(i-1,j)$, $(i+1,j)$, $(i,j-1)$, $(i,j+1)$, except border (and corner) sites which only have 3 (resp. 2) nearest-neighbours. Let A be a stationary random set on G, with proportion $p = P(x \in A)$ and covariance $C(h) = P(x \in A, x+h \in A) - p^2$. Let us introduce some connectivity definitions on the random set A.

Definition 1 *Two sites x and x' of A are said to be connected (denoted by $x \leftrightarrow x'$), if there exists a path linking x to x' entirely contained in A. A path is a sequence of sites $x_1, \ldots, x_n$ where $(x_i, x_{i+1})_{1 \leq i \leq n-1}$ are nearest-neighbours.*

Definition 2 *The connectivity function, τ, is the probability that two sites x and x' of A are connected :*

$$\tau(x,x') = P(x \leftrightarrow x' \mid x, x' \in A) \quad x, x' \in G$$

Lastly, we introduce the set of sites connected to a given one :

Definition 3 *The x_0 connected component, denoted C_{x_0} is the set of sites of A connected to x_0 :*

$$C_{x_0} = \{x \in A \ : \ x \leftrightarrow x_0\}$$

The connected components of A form a partition of A : they are the biggest subsets of A. Let $|.|$ denote the cardinal of any subset of G. In particular, $|G| = n^2$ and $E\{|A|\} = pn^2$.

Our problem is to simulate a random set A with proportion p and covariance $C(h)$, conditionally to the constraints : $\Gamma_1 = (x_l \leftrightarrow x'_l)_{l \in L}$ and $\Gamma_0 = (x_m \not\leftrightarrow x'_m)_{m \in M}$, where M and L are two sets of indices. Γ denotes $\Gamma_0 \cup \Gamma_1$. It must be emphasized that these constraints are not local anymore, but global, since the path linking two points can be anywhere in G. Let $\mathcal{A}(\Gamma)$ be the set of subsets of G satisfying the connectivity constraints Γ : $\mathcal{A}(\Gamma) = \{A \subset G \ : \ \Gamma \text{ conditions are satisfied}\}$. We shall not treat the problem of the consistency of these conditions, and from now on, we shall assume that there exists at least one solution satisfying Γ : $\mathcal{A}(\Gamma) \neq \emptyset$. As the gaussian model is suitable to perform a simulation of a random set $A \in \mathcal{A}(\Gamma)$, we shall use it.

The gaussian random vector $\mathbf{Y} = (Y_x)_{x \in G}$ is associated to G; its mean vector is $E\{Y_x\} = 0$ and its covariance is $\rho(h) = E\{Y_x.Y_{x+h}\}$, $x, x+h \in G$. A realization of $\mathbf{Y}$ is denoted $\mathbf{y} = (y_x)_{x \in G}$. The random set A is defined as the set of sites such that their $\mathbf{Y}$-coordinate are superior to a given threshold y_c :

$$A = \{x \in G \ : \ Y_x \geq y_c\}$$

A is a stationary random set, and $P(x \in A) = p = 1 - G_1(y_c)$ where G_1 stands for the univariate gaussian cdf.

3 Connectivity of non conditional simulations

Before performing conditional simulations, let us illustrate the influence of the model and its parameters on two connectivity characteristics, namely the number of connected components, N, and the connectivity function, τ.

The number of connected components.

From $p = 0$ where $N = 0$, let p increase. For low proportions, new connected components are created in the many empty spaces and $E\{N\}$ increases. But, little by little, different connected components come in contact, tending to decrease $E\{N\}$. These antagonistic effects lead to a balance for a particular proportion where $E\{N\}$ reaches its maximum. Beyond this value, $E\{N\}$ decreases monotonically, and $N = 1$ for $p = 1$.

As the range parameter of the gaussian covariance increases, the regularity of the truncated random set increases too. Hence the connected components are bigger and less numerous. The influence of this parameter is very important, as it can be seen on the left side of figure 1. On the right side of the same figure, two models with the same integral range $\xi = \sum_{h \in \mathbb{Z}^2} C(h)$ are compared : the connected components are less numerous for the spherical model for low proportions, but more numerous for high proportions. These differences are of slight importance compared to a variation of the range parameter.

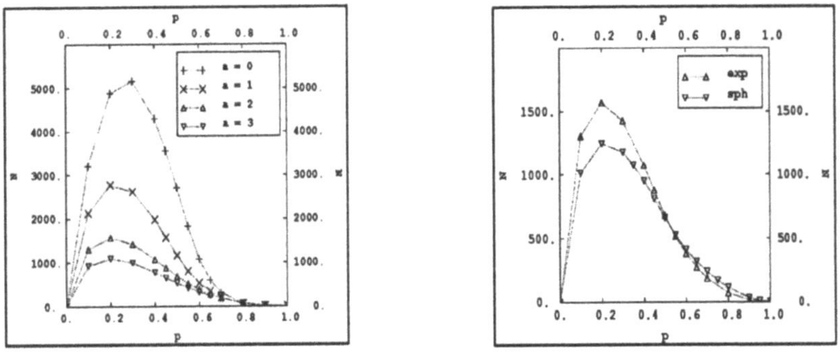

Figure 1: Number of connected components vs. the proportion (mean of 250 simulations on a 200×200 grid). Left : Factorized exponential covariance with parameter $a : \rho(h) = \exp(-a(|h_x|+|h_y|)), \ h = (h_x, h_y) \in \mathbb{Z}^2$. Right : Fact. exp. and spherical covariance with integral range $\xi = 16$.

The connectivity function

For a fixed distance h, the connectivity probability $P_p(x \leftrightarrow x + h \,|\, x, x + h \in A)$ is reported versus the proportion p (figure 2). The existence of a sharp transition separating the domain where $\tau(h) \simeq 0$ from the domain where $\tau(h) \gg 0$ is clearly apparent.

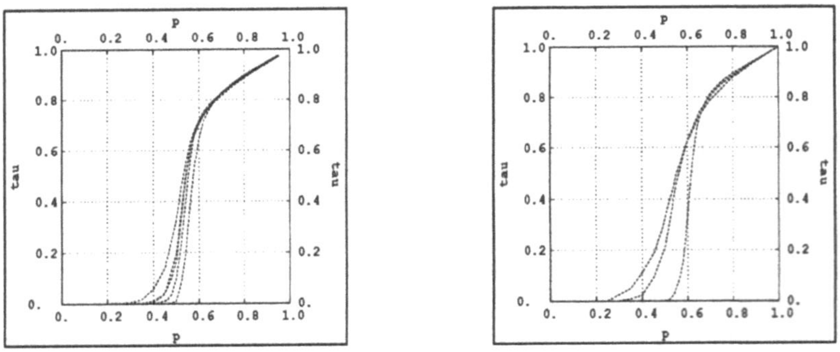

Figure 2: $P_p(x \leftrightarrow x + h \,|\, x, x + h \in A)$ vs. proportion (Fact. exp. covariance). Left : $a = 2$ and $h = (d, d)$, with $d = 10,\ 20,\ 40,\ 80,\ 160$ from left to right. Right : $h = (20, 20)$ with $a = 0,\ 2,\ 4$ from right to left.

On $\mathbb{Z}^2$, it has been shown that the function $\theta(p) = \lim_{|h| \to \infty} P_p(x \leftrightarrow x + h \,|\, x, x + h \in A)$ is such that

$$\theta(p) \quad \begin{cases} = 0 & \text{if } p < p_c \\ > 0 & \text{if } p > p_c \end{cases}$$

if the random set is stationary and has finite range (ie. $\exists R < \infty$ such that if $|h| > R$, then $C(h) = 0$). In the percolation theory, p_c is called the *percolation threshold.* It separates the phenomenon in two distinct phases [8]. If $p < p_c$, all the connected components are finite ; on the contrary, if $p > p_c$, there exists a single infinite connected component on $\mathbb{Z}^2$, denoted C_∞.

It has been observed that the connectivity function $\tau(h) = P(x \leftrightarrow x+h \mid x, x+h \in A)$ is isotropic for the euclidian norm $|h|$, in spite of the covariance and the grid anisotropies. There is no general theoretical justification of this result. However, it has been demonstrated [2] that for uncorrelated bond percolation

$$h \in \mathbb{Q}^2,\ nh \in \mathbb{Z}^2, \quad \tau(nh) \simeq e^{-n.\varphi(p).g(h)} \quad \text{for large } n$$

where $g(h)$ can be extended to the euclidian norm on $\mathbb{R}^2$. The connectivity function follows an exponential decrease. It can be shown [8] that for an ergodic finite range random set the function $\varphi(p)$ in the argument a) is continuous and not decreasing on $]0,1]$, b) $\varphi(p) \to \infty$ for $p \to 0$, c) $\varphi(p) = 0$ when $p > p_c$.

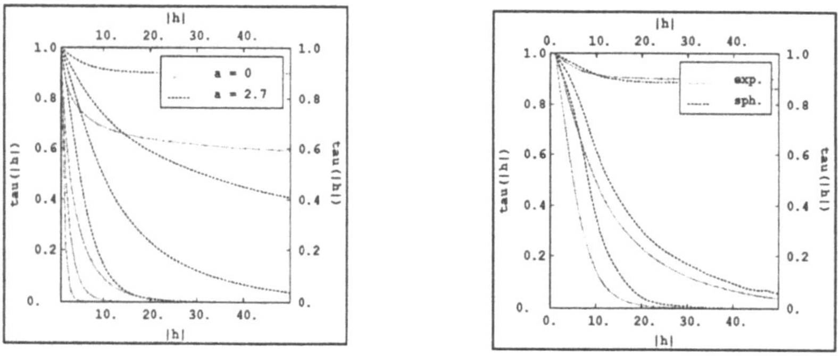

Figure 3: Connectivity function $\tau(|h|)$ Left : Fact. exp. covariance for two ranges ($a = 0$ and $a = 2.7$), and for 4 different proportions (from top to bottom : $p =$ 0.6, 0.5, 0.4, 0.2); Right : Fact. exp. and spherical covariance with same integral range $\xi = 29$, and 3 different proportions (from top to bottom : $p = 0.6,\ 0.4,\ 0.2$)

The connectivity function is illustrated on figure 3. For low proportions, $\tau(h)$ decreases quickly with distance and tends to 0. All the connected components are small and two distant points have an infinitesimal probability to be connected. As p becomes more important, the function $\tau(h)$ decreases more slowly, but it can still be observed that $\tau(h) \to 0$ when $|h| \to \infty$, until the critical value p_c is reached, beyond which $\varphi(p) = 0$. In this case, $\tau(h)$ decreases until it reaches a sill as $|h| \to \infty$. Indeed, for short distances, two connected points of A either belong to a small connected component or belong to the infinite one. But for long distances, two points are connected only if they both belong to the biggest connected component : as $|h| \to \infty$, $\tau(h) = P(x \leftrightarrow x + h \mid x, x + h \in A) \simeq P(x \in C_\infty \text{ and } x + h \in C_\infty) \simeq P(x \in C_\infty)^2$.

On figure 3, it can be seen that $\tau(h)$ increases with the range and is more important in the spherical model than in the factorized exponential model when $p < p_c$, and conversely when $p > p_c$, as the connected components are more regular, hence bigger.

4 Conditional simulation algorithm

Formal Presentation

Let B denote the set of vectors $\mathbf{y} \in I\!R^{|G|}$ satisfying the connectivity conditions Γ. A conditional simulation of $\mathbf{Y}$ is a realization of the conditional probability distribution that respects these constraints. In the truncated gaussian model, the density function to simulate is therefore

$$\mathbf{y} \in I\!R^{|G|} \quad f(\mathbf{y}) = \frac{g(\mathbf{y})\mathbb{1}_B(\mathbf{y})}{\int_{I\!R^{n^2}} g(\mathbf{y})\mathbb{1}_B(\mathbf{y})\,d\mathbf{y}}$$

where g is the multigaussian pdf and $\mathbb{1}$ the indicator function : $\mathbb{1}_B(\mathbf{y}) = 1$ if $\mathbf{y} \in B$; $\mathbb{1}_B(\mathbf{y}) = 0$ otherwise. The set B of solutions has a complex geometry; in particular it is not a product of intervals, since the constraints are not punctual anymore. But, conditionally to $(Y_x = y_x)_{x \neq x_0}$, the set $\tilde{B}_{x_0}$ to which the Y_{x_0} coordinate of $\mathbf{Y}$ must belong, is an interval of one of the three forms : $]-\infty, y_c[$, $[y_c, +\infty[$, $I\!R$.

As it is not possible to simulate $\mathbf{Y} \in B$ with density function f in one step, we use an iterative method : starting with an initial vector $\mathbf{y}_0 \in B$, we modify one of its coordinate at each iteration, until we get a suitable vector. The algorithm we present here is derivated from the Gibbs Sampler, introduced in 1984 by Geman and Geman [7]. A geostatistical application of it can be found in [6].

Let $\{x_1, x_2, \ldots, x_k,\ k \in I\!N\}$ be a visit sequence of G with period $|G| = n^2$. Each period is called a scanning of the vector. The k-th transition from $\mathbf{Y}^k$ to $\mathbf{Y}^{k+1}$ is (denoting $\bar{\mathbf{Y}}^k_{x_k}$ the vector $\mathbf{Y}^k$ without its x_k-coordinate) :

$$Y_x^{k+1} = \begin{cases} Y_x^k & \text{if} \quad x \neq x_k \\ Y_{x_k}^{k+1} \;=\; Y_{x_k}^{SK} + \sigma_{SK}(x_k)U_k & \text{if} \quad x = x_k \end{cases}$$

where $Y_{x_k}^{SK} = Y_{x_k}^{SK}(\bar{\mathbf{Y}}^k_{x^k})$ is the simple kriging of site x_k using all the other sites of G, $\sigma^2_{SK}(x_k)$ is the variance of this kriging, and U_k is an independent (eventually truncated) gaussian variable, such that $Y_{x_k}^{k+1} \in \tilde{B}_{x_k}$. The vectors $\mathbf{Y}^k$ are a Markov chain, since $\mathbf{Y}^k$ is dependent only on the previous state $\mathbf{Y}^{k-1}$, and independent on the others. The transition probability

$$P_k(\mathbf{y}^k, A) = P(\mathbf{Y}^{k+1} \in A \,|\, \mathbf{Y}^k = \mathbf{y}^k) = \int_A dP_k(\mathbf{y}^k, \mathbf{y}^{k+1})$$

is related to the conditional density function of the k-th coordinate $f(.\,|\,\bar{\mathbf{y}}^k_{x_k})$:

$$dP_k(\mathbf{y}^k, \mathbf{v}) = \delta_{\bar{\mathbf{y}}^k_{x_k}}(\bar{\mathbf{v}}_{x_k})\, f(v_{x_k} \,|\, \bar{\mathbf{y}}^k_{x_k})\, dv_{x_k}$$

with

$$f(v_{x_k} \,|\, \bar{\mathbf{y}}^k_{x_k}) = \frac{g_1\left(\dfrac{v_{x_k} - y^{SK}_{x_k}}{\sigma_{SK}(x_k)}\right) \mathbb{1}_{\bar{B}_{x_k}}(v_{x_k})}{\displaystyle\int_{\mathbb{R}} g_1\left(\dfrac{v_{x_k} - y^{SK}_{x_k}}{\sigma_{SK}(x_k)}\right) \mathbb{1}_{\bar{B}_{x_k}}(v_{x_k})\, dv_{x_k}}$$

where g_1 stands for the univariate gaussian pdf. The truncated gaussian probability density function, f, is a stationary probability of P_k [6].This density is unique if B is compact and connected [4], or if $B = \mathbb{R}^m$ [9]. If $B \neq \mathbb{R}^m$ is connected but not compact, to our knowledge, there is no result concerning the uniqueness of the stationary probability. But an approximation can be made. Let us consider the truncated gaussian model on the set $B \cap [-a, a]^{n^2}$ where a is any positive, large enough, constant. The model is modified —we do not consider the gaussian model anymore, but the truncated gaussian model on the compact set $[-a, a]^{n^2}$— but in practice, $a > 10$ has only slight changes on the model.

It is not always possible to determine if B is a connected set of solutions, or not. However, in some simple cases, B is always connected :

- If $\Gamma = \Gamma_1$, ie. all constraints are connection constraints, then $[y_c, +\infty[^{n^2} \subset B$, hence B is connected.
- Conversely, if $\Gamma = \Gamma_0$, ie. all constraints are non-connection constraints, then $]-\infty, y_c[^{n^2} \subset B$, hence B is connected.
- If $\Gamma = \Gamma_0 \cup \Gamma_1$, but in each non-connection constraint of Γ_0, at least one of the two connected components has one conditioning point only, then B is connected. A conditioning point is one of the points x_m or x'_m, $m \in M$ seen in the definition of Γ_0 (see section 2).

For this three simple cases, it is true that the Gibbs sampler samples all the solutions. In the other cases, it is not always true.

In the other cases

Coming back to $\mathcal{A}(\Gamma)$, the set of subsets of G satisfying the connectivity constraints Γ, we build an elementary transformation at site x, denoted t_x, which transforms a solution $\mathbf{x} \in \mathcal{A}(\Gamma)$ into another solution $\mathbf{x}' \in \mathcal{A}(\Gamma)$. We define t_x as :

$$t_x(\mathbf{x}) = \begin{cases} \begin{cases} \mathbf{x} + \{x\} & \text{if } x \notin \mathbf{x} \\ \mathbf{x} \backslash \{x\} & \text{if } x \in \mathbf{x} \end{cases} & \text{if } t_x(\mathbf{x}) \in \mathcal{A}(\Gamma) \\ \mathbf{x} \quad \text{otherwise} & \end{cases}$$

This elementary transformation defines a deterministic dynamic on $\mathcal{A}(\Gamma)$ which is equivalent to the Gibbs sampler. Now, we build a relation, denoted $\rightleftarrows$, and we say that $\mathbf{x} \rightleftarrows \mathbf{x}'$ ($\mathbf{x}$ is related to $\mathbf{x}'$) if there exists a finite sequence of elementary transformations from $\mathbf{x}$ to $\mathbf{x}'$ (and conversely). Clearly, $\rightleftarrows$ is an equivalence relation. The associated equivalence classes are distinct classes of solutions for the constraints Γ, when using the Gibbs sampler. It must be emphasized that it is not possible to link two solutions belonging to two distinct classes using elementary transformations. An example is given figure 4. The three relations between the 4 conditioning points $\{x_1, \dots, x_4\}$ are : $(x_1 \leftrightarrow x_2)$; $(x_3 \leftrightarrow x_4)$; $(x_1 \not\leftrightarrow x_3)$. There exists 3 classes of solutions, depicted on this figure, and it is impossible to go from one class to an other using elementary transformations, without breaking one of the 3 connectivity constraints.

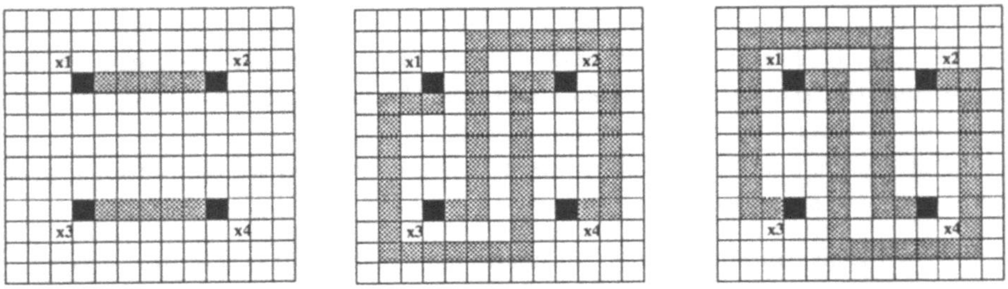

Figure 4: Conditions are : $(x_1 \leftrightarrow x_2)$; $(x_3 \leftrightarrow x_4)$; $(x_1 \not\leftrightarrow x_3)$. The three solutions belong to three distinct equivalence classes.

More generally, the consequence on the algorithm is that, starting from an initial image $\mathbf{x}_0$, the Gibbs sampler covers only the solutions belonging to the same class as $\mathbf{x}_0$, ignoring all the others. Let $(B_i)_{1 \leq i \leq n_c}$ be the classes of vectors $\mathbf{Y}$ corresponding to the n_c classes of solutions $\mathcal{A}_i(\Gamma)$, and $g(B_i)$ be their measure for the gaussian density g, $(g(\mathbb{R}^{n^2}) = 1)$. Then, f, the truncated gaussian density on B becomes

$$f(\mathbf{y}) = \frac{g(\mathbf{y})\mathbb{1}_B(\mathbf{y})}{g(B)} = \frac{g(\mathbf{y})\sum_i \mathbb{1}_{B_i}(\mathbf{y})}{\sum_i g(B_i)} = \sum_i \frac{g(B_i)}{\sum_i g(B_i)} f_i(\mathbf{y})$$

where

$$f_i(\mathbf{y}) = \frac{g(\mathbf{y})\mathbb{1}_{B_i}(\mathbf{y})}{g(B_i)}$$

is the truncated gaussian density on B_i. The f_i are the densities of a stationary law for the transition probability P_k. f is a weighted sum of the f_i, where the weights are given by the probability system $g(B_i)/g(B)$.

In summary, if there are several classes of solutions, the Gibbs Sampler will not explore all the solutions of B, but only the ones belonging to the class of the initial image $\mathbf{x}_0$. Furthermore, the simulated density is not f anymore, but f_i, the truncated gaussian density on the class B_i.

Description of the algorithm

In all cases, the algorithm is the following :

1) Build $\boldsymbol{x}_0 \in \mathcal{A}(\Gamma)$ and then build $\boldsymbol{y}_0 \in B$

2) Visit all sites of G, following a sequence $(x_k)_{k\in I\!N}$ with period $|G| = n^2$; at the k-th iteration :

a) Compute simple kriging of site x_k, $Y_{x_k}^{SK}$, and its kriging variance $\sigma_{SK}^2(x_k)$.

b) Compute the interval ($]-\infty, y_c[$, $[y_c, +\infty[$ or $I\!R$) which the x_k coordinate of vector $\boldsymbol{Y}$ must belong to, in order to respect the constraints Γ.

c) Sample an independent gaussian variable U_k belonging to $]-\infty, u_c[$, $[u_c, +\infty[$ or $I\!R$, following a), where $u_c = \dfrac{y_c - Y_{x_k}^{SK}}{\sigma_{SK}(x_k)}$.

d) Assign the value $Y_{x_k}^{SK} + \sigma_{SK}(x_K)U_k \rightarrow Y_{x_k}$.

3) Stop iterations.

Remark : In practice, $\mathbf{y}_0$ is built using a non-correlated gaussian vector.

5 Implementation

Treating a few examples will lead us to consider the following problems : the consistency between the model of gaussian random vector and the connectivity constraints, the speed of convergence and the existence of several classes of solutions.

Influence of the conditioning

In this paragraph, we assume that the connectivity constraints have a unique solution class, B. In this case, for all initial vectors $\mathbf{y}_0 \in B$, $(\mathbf{Y}^k)_{k\in I\!N}$ converge weakly to $\mathbf{Y}$, the truncated gaussian vector on B with density $f(\mathbf{y}) = g(\mathbf{y})1\!\!1_B(\mathbf{y})/g(B)$. This random vector is not gaussian anymore; in particular, its first and second moment statistics do not characterize the entire law, and these moments are not explicitly calculable. If the conditioning constraints are sampled from a non conditional gaussian simulation, $\mathbf{Y}$ is gaussian again. But, we have not worked in this direction, as we want to illustrate the influence of non-gaussian constraints.

Suppose that Γ is made up of a unique condition : " $x \leftrightarrow x'$ ". The measure of the set of solutions is just the connectivity function : $g(B) = P(x \leftrightarrow x') = \tau(x' - x)$. On figure 3, we have seen that this function, almost equal to zero if $p < 0.4$, increases sharply around $p = 0.5$. Now, if Γ is made of several connection and non-connection constraints, the probability that the configuration is verified in the non-conditional model, $P(\Gamma \text{ is verified }) = g(B)$, has a maximum [1] for a proportion which is a compromise solution between connection and non-connection constraints. Experimentally, this proportion is close to the percolation threshold; furthermore, for this value of the proportion, the variability of the connectivity characteristics (connectivity function, number of connected components, size and shape of them etc.) is maximum. So, it can be forecast that the simulated model, f, will be different below, around or above the percolation threshold. Let us illustrate this on an example. For the connectivity conditions depicted here —denoted Γ'—, simulations are performed for three values of the theoretical proportion : inferior, close and superior to p_c ($p = 0.2$, $p = 0.4$, $p = 0.7$). The following experimental statistics are calculated in function of the number k of scannings of G : ${\sigma^{\star}}^2(k)$, $m^\star(k)$, $\rho^\star(k)$ and $p^\star(k)$, respectively equal to the experimental variance, mean, covariance between nearest neighbours and proportion of the vector $\mathbf{y}^k$.

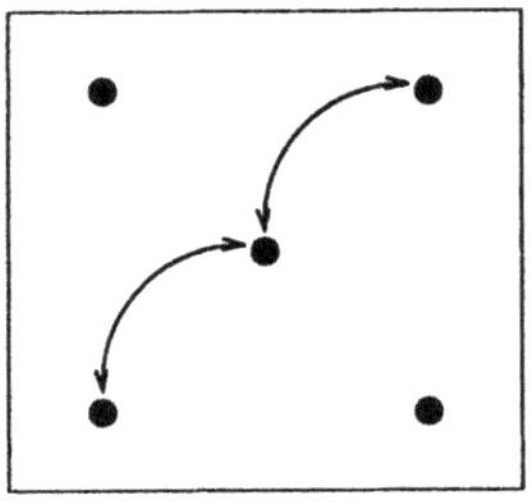

If the proportion is close to the percolation threshold, the parameters of the non-conditional model are well respected. On the contrary, if the proportion is small compared to the percolation threshold, the probability of the connection constraints is low. The number of sites above the threshold y_c increases in order to respect the connectivity; hence $p^\star \geq p$ and $E\{(Y_x)^k\} \geq 0$. Conversely, if the theoretical proportion is important compared to p_c, the non-connection constraints have a low probability, hence $p^\star \leq p$ and $E\{(Y_x)^k\} \leq 0$. The variogram at iteration 100 is well reproduced for $p = 0.4$ and $p = 0.7$, but the sill is superior to 1 for $p = 0.2$. The histogram is shifted upwards to higher values for $p = 0.2$ and downwards to lower values for $p = 0.7$, as can be seen from the mean given earlier.

In the general case, the lower the probability of the constraints in the non-conditional gaussian random vector is, the more the statistics calculated on the conditional simulations differ from those of the model. In fact, in order to respect them, connectivity information should be integrated in the inference of the model.

[1]This maximum exists : $g(B) = 0$ for $p = 0$ and for $p = 1$; but $g(B) > 0$ if $p \in]0, 1[$.

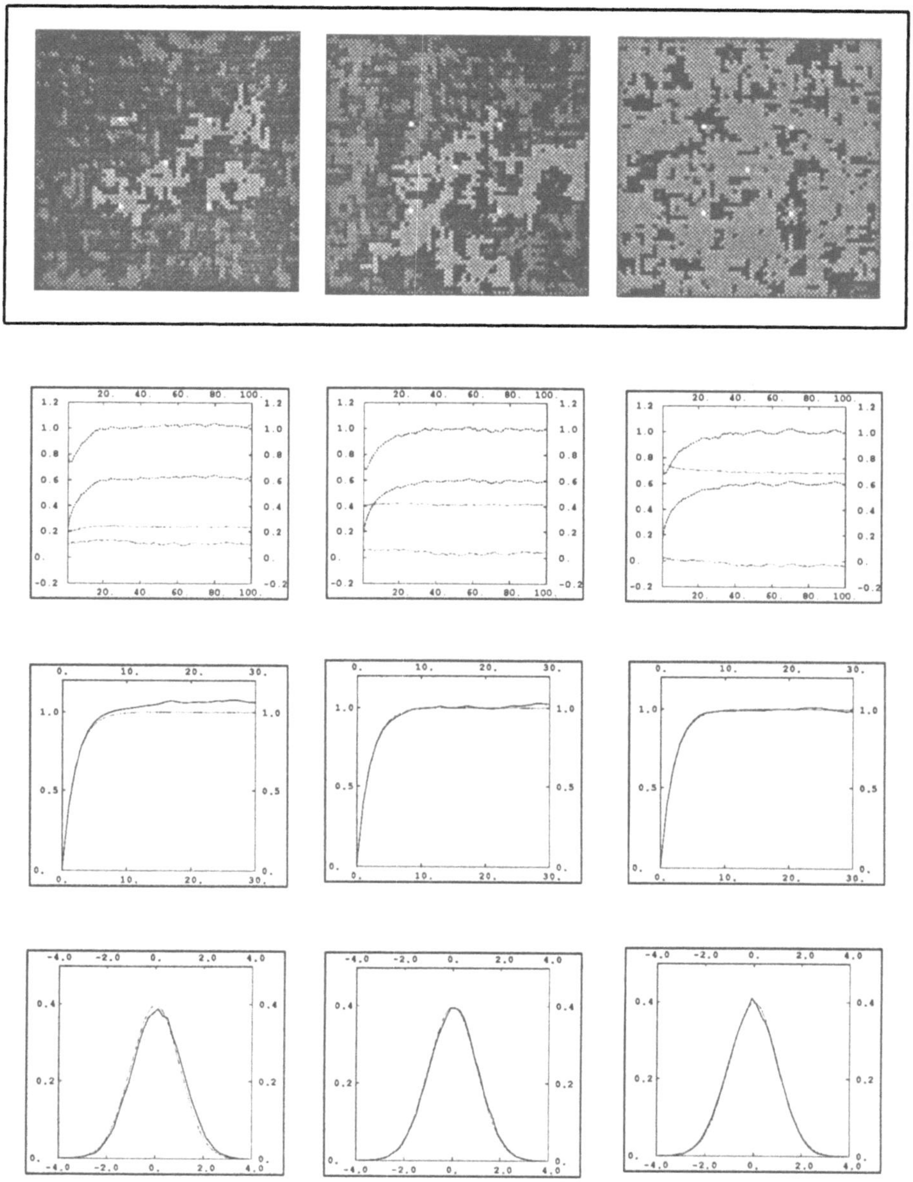

Figure 5: Top : 3 conditional simulations for the constraints Γ' : $p = 0.2$; $p = 0.4$ and $p = 0.7$. Pores are black; grains are grey (light grey for the 3 conditioning connected components). The 5 conditioning points are white. Then : elementary statistics $\sigma^{\star^2}(k)$, $m^\star(k)$, $\rho^\star(k)$ and $p^\star(k)$ (mean of 12 simulations) vs. the number of scannings, variogram and histogram at 100-th scanning.

Speed of convergence

It can be observed (figure 5) that the 4 experimental statistics reach a sill together. The stabilization of the experimental variance, which has been chosen as the indicator of the convergence depends on the model of covariance, and on the simulations constraints : proportion p and connectivity constraints. The range of the covariance is a very important factor. For non-conditional simulations, the number of scannings necessary to have a good stabilization increases with the range, as reported in Table 1.

Table 1

a	ρ	k
2.0	0.605	40
3.0	0.715	100
4.0	0.779	>400

This table shows that the Gibbs sampler should not be used for non conditional simulations, because it is too slow. But, the stronger the connectivity constraints are, the faster the algorithm converges. This can be seen on figure 6 where the experimental variance is reported vs. the number of scannings, for three values of the proportion and a non-conditional simulation. $p = 0.2$ is the most improbable constraint, thus the strongest : it converges the most rapidly.

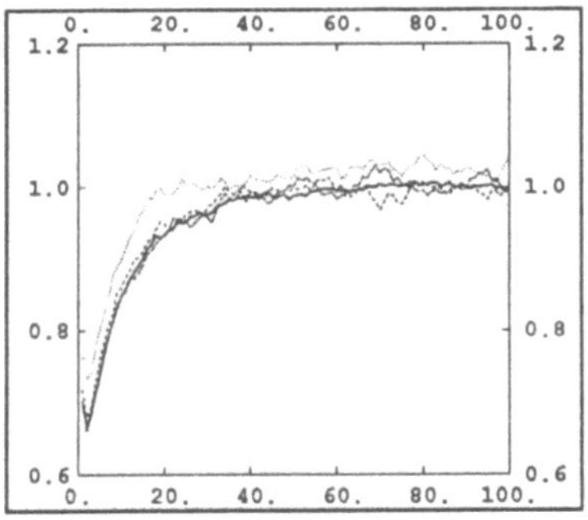

Figure 6: Experimental variance vs. number of scannings; mean of 12 conditional simulations with constraints Γ' : $\cdots$ $p = 0.2$; - - - $p = 0.4$; -·-·- $p = 0.7$; — mean of 12 non conditional simulations.

Several classes of solutions

If there are several classes of solutions $(B_i)_{1 \leq i \leq n_c}$ with probability $g(B_i)$, the initial image will determine the class to which the simulations belong. In order to simulate the density f on B, the initial image should be randomly chosen in the i-th class with the probability $g(B_i)/g(B)$. Unfortunately, these probabilities are not calculable. Two conditional simulations with the constraints defined figure 4, belonging respectively to the first and second class of solutions, are depicted figure 7 with their elementary statistics. For the first class of solutions, the model is well respected; on the contrary, the experimental mean and proportion are very different of the model for the second one, which is more unlikely in the gaussian model.

The connectivity constraints do not always specify completely the model to simulate when using the Gibbs sampler. An arbitrary decision remains in the choice of the initial image, or in their weights. But this arbitrariness is also a flexibility of the method, as it gives to the geologist the possibility of integrating qualitative interpretations.

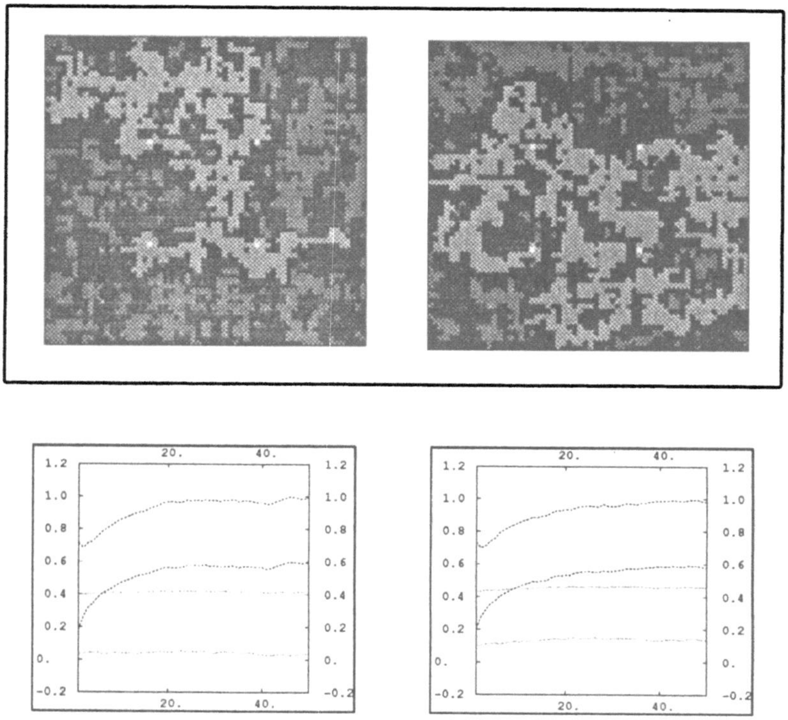

Figure 7: Conditional simulations (top) and elementary statistics $\sigma^{\star 2}(k)$, $m^\star(k)$, $\rho^\star(k)$ and $p^\star(k)$ (mean of 12 simulations) for two classes of solutions for the same connectivity constraints (defined figure 4).

6 Conclusion

The Gibbs sampler is a powerful and flexible algorithm, which allows to perform conditional simulations integrating complex constraints. In this paper, lithofacies are simulated conditionally to connectivity constraints. But, the stronger the conditioning is, the more different the simulated density and the model are. Therefore, statistics on conditional simulations can differ notably from those of the model. In order to respect them, the inference of the model should integrate the connectivity constraints. It may happen that the conditioning leads to several classes of solutions. In that case, conditional simulations will belong to the same class as the initial image, which has to be chosen carefully. Last, for the moment, the algorithm is quite slow, requiring numerous connected component computations. In its current version, a 100×100 grid with 50 scannings needs about 2 hours, on a usual workstation.

References

[1] F. Alabert. *Stochastic imaging of spatial distributions using hard and soft data.* Msc. thesis, Stanford U, 1987.

[2] K.S. Alexander et al. The Wulff construction and asymptotics of the finite cluster distribution for two dimensional Bernouilli percolation. *Comm. Math. Phys.*, 1989.

[3] Cl. Deutsch. Conditionning reservoir models to well test information. In A. Soares, editor, *Geostatistics TRÓIA '92*, volume 1, pages 505–518, Dordrecht Holland, 1993. Kluwer Academic Press.

[4] J. L. Doob. *Stochastic processes.* John Wiley & Sons, Inc., New York, 1953.

[5] X. Freulon. Conditional simulation of a gaussian random vector with non-linear and/or noisy observations. In M. Armstrong, editor, *Workshop on geostatistical simulations (to appear)*, Dordrecht Holland, 1994. Kluwer Academic Press.

[6] X. Freulon and Ch. de Fouquet. Conditioning a gaussian model with inequalities. In A. Soares, editor, *Geostatistics TRÓIA '92*, volume 1, pages 201–212, Dordrecht Holland, 1993. Kluwer Academic Press.

[7] S. Geman and D. Geman. Stochastic relaxation, Gibbs distribution and the Bayesian restoration of images. *I.E.E.E. transactions : Pattern Analysis and Machine Intelligence*, 6:721–741, 1984.

[8] G. Grimmet. *Percolation.* Springer-Verlag, New York, 1992.

[9] X. Guyon. *Champs aléatoires sur un réseau.* Masson, Paris, 1993.

[10] A.G. Journel and F.G. Alabert. Focusing on spatial connectivity of extreme-valued attributes: Stochastic indicator models of reservoir heterogeneities. *SPE, 63rd Annual Conference, Huston, TX*, 1988.

[11] G. Matheron et al. Conditional simulation of the geometry of fluviodeltaic reservoirs. *SPE, 62nd Annual Conference, Dallas, TX*, 1987.

[12] G. Verly. Sequential gaussian co-simulation: A simulation method integrating several types of information. In A. Soares, editor, *Geostatistics TRÓIA'92*, volume 1, pages 543–554, Dordrecht Holland, 1993. Kluwer Academic Press.

DISCUSSION FOLLOWING SESSION NO 6

Chairman: Henning Omre

Papers: Jaquet; Allard

After Olivier Jaquet's paper, the chairman thanked him for a challenging talk with forward looking questions.

Jeulin — You want to simulate a network with strong global connectivity properties, and yet you used a local approach by simulating small blocks. Consequently there are major problems relating the local connectivity to the global connectivity. I would like to suggest a global rather than local development of a network using some kind of random walk approach, for example by using lattice gas models. The random walks could have branching processes to give interconnected random walks. In this type of approach you can also include the fact that you have blocks of different strengths, and others that the galleries cannot go through.

Jaquet — They could also be blocks that are in some way reflect the geology. We could constrain the density of the network in space.

Jeulin — In your diagram of the karstic system, you can see some sort of tesselation in space that must have a geolcgical meaning.

Jaquet — That is where I could include my geological information. So it would be global in terms of connectivity and local in terms of blocks.

Daly — I am rather worried about the sampling strategy. Speleologists cannot get through a tunnel less than 30 cm wide, and yet it might well widen out just a little further on. So it is clearly very difficult to sample the system properly without introducing a bias. Would it be possible to analyse some 2D slices and use some sort of stereological procedure to properly estimate your variograms in 3D?

Jaquet — As you say, there is a serious bias in the data. I have got all the data in 3D:– the length, the orientation and the diameters. Inferring the unknown conduits is a real problem. I need a way to include them in the model.

Armstrong — You say that your model is not perfect because the conduits do not connect up. At first that sounds quite right. But Colin Daly is saying that

M. Armstrong and P. A. Dowd (eds.), Geostatistical Simulations, 212–215.

the 30 cm conduits (the only ones you can see) do not join up but there may well be smaller ones that do connect these. May be your picture is not so silly, given that it only shows the big ones.

Jaquet — But I need to include the smaller ones as well.

Armstrong — Yes but the fact that things are not connected at 30 cms does not mean that they are not connected at all.

Omre — Modelling of fractures, faults and conduits in reservoirs is one of the things that have been really underworked in petroleum. The problems you are working on, are being faced by lots of people. Several current EEC projects in petroleum are on the modelling of faults in reservoirs. Much of the work is being done as marked point processes rather than on a grid, because it is more flexible. So we are not forcing the geometry. This is a field of ongoing research.

Jaquet — For the modelling of flow within fractured rocks we are using a code called NAPSAC, which allows us to generate fracture networks. Then the fracture connectivity is determined., and the effective permeability is computed. This approach only considers the flow through the fractures. Flow through the matrix elements is neglected on the grounds of computer efficiency.

Omre — The challenge is to bring the geological model into the simulator.

Jaquet — The hydrogeologist wants to make transport simulations but they cannot match their results to the observed response because they have no idea of the networkgeometry.

The chairman then stopped the discussions to allow the next presentation to go on.

Haas — I think it is very interesting to be able to take account of the constraints but the difficulty is that these constraints can be completely unrealistic. For example, if you try to connect two wells and if the mean proportion is very low, you will obtain an image where all the good permeability values will lie between the two wells and nothing outside.

Allard — That is the problem of the consistency between the model and the constraints.

Haas — My conclusion is that the connectivity must be respected in probability and not strictly.

Chairman — Although the two papers are quite different, they both have the similarity of respecting constraints. The general discussion should focus on living up to complex constraints.

Armstrong (to Denis Allard). In your first figure, all of the connectivity lies between the 3 wells, because of the low probability. If you had dropped the probability even lower to say 0.1, thereby increasing the incompatibility between the constraints and the model to an absurd point, what does your method do? Does it explode? What happens?

Allard No it does not explode. It always give an image, but it can be very different from the model you are expecting. There is always at least one possible path but if you give too little probability, then two things happen. Firstly, most sites will be very near the constraint and the variogram will not be respected and secondly, the experimental proportion will be higher than the input value. For example if you input a proportion of 0.1, you could get an image with proportion 0.2, which is completely unrealistic.

Omre Are there problems finding an initial state ?

Allard The initial state is given by hand.

Jeulin You showed earlier that the connectivity constraints are local ones, whereas the model constraints are global. So you do not work on the same scale. It is very difficult to respect constraints at different scales. You can have constraints with a very low probability of occurrence.

Allard That is why the connectivity function is interesting. It gives you an idea of the probability of the constraint. The connectivity function is a 2 point function. There are rules for calculating the bounds for the probabilities of 3 (or 4 or 5) points being connected. As you know, the connectivity function is the geometric covariogram of a typical component of the random set. So it has certain general inequality properties, which can be used to establish upper and lower bounds.

Jeulin If you have mixed conditions, it is not difficult to calculate it from just the equality conditions, because you have simple relationships. For example, if you want one point in one component and a second in another, you can use a closure relation.

Allard You can get interesting results using 2 inequalities that are FKG and BK theorems.

Daly (to Hegstad et al.). Have you thought of using simulated annealing to choose which of the connected subsets? I would be most interested to see how this could be done.

Omre In Hegstad's paper you distinguish between your model (which is valid for many reservoirs of that type). Then you have some constraints that are reservoir specific, like the one you have. The model is general. The reservoir specific constraints will be the B function and the temperature. In the initial stages, you have no loose constraints. Then you start iterating and tightening in the constraints.

Daly How do you tighten in a constraint ?

Omre The two points should be connected. It is a 0−1. At the end you tighten down by reducing the T and you end up with the optimum uniformly over all solutions. So you start out with very loose constraints and you tighten up by cooling down.

Allard How do you calculate the distance between what you expect and what you get ?

Omre The distance is as T goes down to 0. It does not matter because in the end it will be a direct function or a function having only positive probability. You can use any function because you can cool it down mathematically to zero, but in practice you do not cool it all the way down. So you should use a suitable function; for example a Gibb's sampler.

Allard In my simulations, either the constraints are respected and the objective function equals 1, or they are not and it equals 0. So how could you integrate the connectivity in a simulated annealing algorithm when the objective function takes only two values ?

Omre I see your point; it may be more complex than I had previously described.

Allard That is why I used the Gibb's sampler with an initial image that already respects the constraints.

THE PROS AND CONS OF THE TRUNCATED GAUSSIAN METHOD

Galli A.[1], Beucher H.[1], Le Loc'h G.[1], and Doligez B[2].
and Heresim Group
(1)Centre de Géostatistique, Ecole des Mines de Paris
35 rue St–Honoré, 77305 Fontainebleau, France

(2) Institut Français du Pétrole
1–4 Avenue de Bois–Préau
BP 311
92506 Rueil Malmaison Cedex, France

Abstract

The truncated gaussian method [6][8] has been widely used in the past in order to simulate indicators. In this paper we investigate the potential and limitations of this method, from a geostatistical point of view and also from a practical one.
First we present the method in detail including the hypotheses behind it. In particular we show that this method ensures consistency for the model in terms of the indicators variograms and cross variograms. Examples of the types of variograms that can and cannot be fitted using this method are presented.
Then we discuss how to take into account vertical non stationarity for the indicators, as this seems to be the usual case (due to sequential evolutions in a sedimentary unit). This point will be illustrated via several examples. We then explain how the method can deal with a complex 3D non stationarity, and discuss the practical problems of using it in this case.
Finally we show that even though this method was initially designed for lithofacies presenting sequences (when more than two facies are present), it can be extended to some other cases.

INTRODUCTION

Because of increasing interest in the petroleum industry in reservoir characterization, many stochastic models have been developed over the past few years. Among these is the truncated gaussian, popularized under the name HERESIM (a joint group between the Institut Français du Pétrole and the Centre de Géostatistique de l'Ecole des Mines de Paris). It plays an important role because:

M. Armstrong and P. A. Dowd (eds.), Geostatistical Simulations, 217–233.

- the algorithm is easy to use, fast, flexible and easy to constrain with external information like proportions;
- the vertical proportion curves, which were integrated into this algorithm by geostatisticians for modeling purposes, have proved to be a very simple but yet powerful tool for geologists for sequence stratigraphy.

In this paper we will discuss in detail the mathematical aspects of truncated gaussians, starting with some basic facts concerning indicators. As a consequence of these we use a completely specified model for indicators using truncated gaussians, as will be shown in the next sections. The relationships between the variogram of the gaussian random function and its indicator variograms are presented in detail for the stationary case. The nonstationary case which appears to be one of the strengths of this method is also discussed. Finally we present some extensions of the initial method that allow us to break sequences into sets of subsequences and to model indicators with different anisotropies.

BASIC PROPERTIES OF INDICATORS

The aim of this section is to review some elementary properties of indicators that are so well known that most of them have never been put in writing explicitly.

Let F be a random set (in this context sets are mainly lithofacies, or classes defined on one or some continuous functions), the indicator of this set $1_F(x)$ is the random function taking the value 1 if x is in F or zero otherwise. Let $\overline{F}$ denote the complement of F.

- $1_F(x) = 1 - 1_{\overline{F}}(x)$
- $1_{F \cup \overline{F}}(x) = 1 = 1_F(x) + 1_{\overline{F}}(x)$
- If we let $P_F(x) = E(1_F(x))$, we have $0 \leqslant P_F(x) \leqslant 1$ and $P_F(x) + P_{\overline{F}}(x) = 1$
- The variance of $1_F(x)$ is $P_F(x)(1 - P_F(x)) \leqslant 0.25$
- The non centered covariance is :
  $C_F(x,y) = E(1_F(x)\, 1_F(y)) = P(x \in F \text{ and } y \in F)$
- If we let $\sigma_F(x,y)$ and $\gamma_F(x,y)$ be respectively the centered covariance and the variogram of $1_F(x)$ we have the following property :
  $0 \leq \gamma_F(x,y) \leq 0.5$

So an indicator variogram cannot be modeled by a power variogram (even in the intrinsic case)

Proposition : If F is a stationary random set then:

$$\gamma_F(h) = \gamma_{\overline{F}}(h) \tag{1}$$

$$\sigma_F(h) = \sigma_{\overline{F}}(h) = -\sigma_{F\overline{F}}(h) = -\sigma_{\overline{F}F}(h) \tag{2}$$

(see annex for demonstration).

The practical consequences are that :

- two complementary random sets are correlated and their cross covariance is the simple covariance up to a sign,
- the indicator covariance (or variogram) is far from being a connectivity index, because if it were, a set and its background would have the same connectivity which is clearly not true.

From (2) we can easily obtain the relationship for the case of 3 facies (sets) A,B,C.

Proposition : If A, B and C are stationary random sets such that $A \cup B \cup C = R^n$, then:

$$\begin{aligned}\sigma_A &= -\sigma_{AB} - \sigma_{AC}\\ \sigma_B &= -\sigma_{BA} - \sigma_{BC}\\ \sigma_C &= -\sigma_{CA} - \sigma_{CB}\end{aligned} \tag{3}$$

$$\sigma_{AB}(h) - \sigma_{BA}(h) = \sigma_{BC}(h) - \sigma_{CB}(h) = \sigma_{CA}(h) - \sigma_{AC}(h)$$

(see annex for demonstration).

As A, B and C are a partition of space, they cannot be all uncorrelated. Furthermore we will show that it is impossible for two facies not to be correlated. Assume we have n facies, two of which A and B are uncorrelated. We call C the union of the remaining $n-2$ facies.
From (3) we obtain the relation

$$\sigma_C(h) = \sigma_A(h) + \sigma_B(h)$$

Keeping in mind that $\sigma(0) = P(1-P)$ we find :

$$P_C(1-P_C) = P_A(1-P_A) + P_B(1-P_B)$$

which implies $P_A P_B = 0$; that is, one of the facies A or B is not present, which contradicts the hypothesis.
Until now we have been discussing consistency relations when working with different indicators, but there are also drastic conditions for a covariance to be the covariance of an indicator (e.g. more than being a positive definite function) .

Proposition [7]: The variogram of an indicator must satisfy the triangle inequality:

$$\gamma(a+b) \leqslant \gamma(a) + \gamma(b)$$

As a trivial consequence of this (necessary but not sufficient condition, see again [7]) we see that for small distances the behavior of an indicator variogram is in $|h|^\alpha$ with $\alpha \leqslant 1$.

It is now clear that fitting a consistent model for even a few indicators is almost impossible without completely specifying the random sets (see [9] for a general discussion about types of models)

The truncated gaussian model is one of the simplest complete models for indicators, which furthermore makes simulations easy to perform.

THE TRUNCATED GAUSSIAN MODEL

In this model the different facies (sets) are obtained by truncating a stationary multigaussian random function Y(x) with zero mean and unit variance – x being a point in R^3.

Assuming we want to model n facies $F_1, F_2,..., F_n$, we define:

$$F_i = \{x \in R^3 \ ; \ s_{i-1} < Y(x) \leqslant s_i\} \tag{4}$$

$$P_i = E(1_{Fi}) = P(s_{i-1} < Y(x) \leqslant s_i) = G(s_i) - G(s_{i-1})$$

with P_i being the proportion of facies F_i, and G the cumulative normal distribution. The s_i are called thresholds and they satisify

$$s_0 = -\infty \quad , \ s_n = +\infty$$

$$s_i = G^{-1} \left(\sum_{j=1}^{i} P_j \right) \Leftrightarrow P_i = G(s_i) - G(s_{i-1}) \tag{5}$$

First properties of this model

From (4) we see that in the continuous case only facies F_{i-1} and F_{i+1} can be contiguous to facies F_i. In practice the simulation is performed on the nodes of a regular grid, so that we can have transitions between facies i and the other facies. However this model has been designed to be (and in fact is) compatible with practical situations in which an ordering can be defined with high transition probabilities between facies i−1 and i, i and i+1.

From (5) we see that the thresholds are in bijection with proportions of facies. Knowing the proportion of each facies we can compute these thresholds so that after truncating the gaussian Y(x) we respect the actual proportion of these facies.

Indicator simple and cross covariances

$$C_{F_iF_j}(x \ , \ x+h) = P(x \in F_i \text{ and } x+h \in F_j)$$

From the definition of F_i and F_j this is:

$$P\,(\,s_{i-1} < Y(x) \leqslant s_i \;\text{ and }\; s_{j-1} < Y(x+h) \leqslant s_j\,)$$

As Y(x) is multigaussian, we find:

$$C_{F_iF_j}(x, x+h) = \frac{1}{2\pi\sqrt{1-\varrho^2}} \int_{s_{i-1}}^{s_i} \int_{s_{j-1}}^{s_j} e^{-[\frac{u^2+v^2-2\varrho uv}{2(1-\varrho^2)}]}\, du\, dv \qquad (6)$$

where ϱ is the covariance of Y(x) for distance h.
Equation (6) shows that the indicator cross covariances are determined by the covariance ϱ, the thresholds (that is proportions and orders of facies). Using (6) automatically insures consistency for the total set of simple and cross covariances for the indicators. Before considering the variation of $C_{F_iF_j}$ with these parameters in more detail, we make another important comment. In formula (6) the integrand depends only on h through $\varrho(h)$, so that if the thresholds, (that is the proportions) are independent of the location then $C_{F_iF_j}$ depends only on h, and the facies are stationary. However formula (6) allows us to model covariances of non−stationary indicators, through variations of proportions in space (and still in a consistent way) .
Classically formula (6) can be expanded in Hermite polynomials [5] [1] . But for both analytic and practical studies it is better to start with another formula developed by G. Matheron.

<u>Proposition</u> [5]
Let F = {x ; Y(x) $\leqslant$ s}, where Y is multigaussian, with zero mean, unit variance and covariance ϱ:

$$\sigma_F = \frac{1}{2\pi} \int_0^{\varrho} e^{-\frac{s^2}{1+x}} \frac{dx}{\sqrt{1-x^2}} \qquad (7)$$

Remark: it is easy to derive an analoguous formula for (6) involving only one integral, although not so simple (4 terms).

Behaviour of indicator variogram

Assuming constant thresholds, with the lowest one equal to $-\infty$ (first facies in the ordering), we see first from (7) that $\sigma_F(h)$ is symmetrical in s. This means that the indicator covariance is the same for proportion p $\leqslant$ 50% and proportion 1−p (see (2)). It can be shown [5] that near the origin the behaviour of σ_F is in $\sqrt{1-\varrho^2}$. Differentiating (7) with respect to ϱ shows that the indicator covariance is an increasing function of ϱ. On figure 1 we can see this behaviour (standardized covariance) for proportions from 10% to 50%.
Differentiating the standardized centered covariance $\tilde{\sigma}_F$ with respect to s we find:

$$\frac{\partial}{\partial s}\,\bar{\sigma}_F = K \int_0^{\varrho} I(x,s)\, e^{-\frac{s^2}{1+x}}\, \frac{dx}{\sqrt{1-x^2}}$$

$$\text{with} \quad K = \frac{1}{2n\,[G(a)\,(1-G(a))]^2}$$

$$I(x,s) = \frac{-2s}{1+x}\, G(s)\,(1-G(s)) - g(s)\,(1-2G(s))$$

For $s<0$ using $s\,G(s) \geqslant -g(s)$ we see that $I \geq 0$, and similarly for the integral. This shows that the range is an increasing function of the proportion up to 50%. This evolution is shown Figure 2, where the curves indicate from the lower to the upper the indicator variogram, of first facies with proportions 5%, 10%, 20%, 30%, 50%. On Figure 3 going from the bottom to the top we see the destructuration of the standardized variogram when a facies with proportion 10% comes after a facies having cumulated proportions of 0%, 10%, 20%, 30%, 40%.

To sum up : In the case of constant proportions the indicators are stationary, the behaviour of their variograms near the origin is fixed by ϱ, their shapes depend on the proportions, and the ordering. Their ranges are lower than the range of ϱ. However they are monotonic functions of $\varrho(h)$. This means they have the same overall aspect and anisotropy. In particular, even if we have some freedom it is impossible to model two facies having different anisotropies using the truncated gaussian with constant proportions.

Non-constant proportions

This is the most frequent case, at least when modeling sedimentary reservoirs. This is strongly linked with cycles, controlled by relative sea level, giving a vertical evolution characterized by vertical proportion curves. See Figures 4 and 5 for examples of vertical proportions and resulting simulations honoring this evolution. Variations of proportions in the horizontal direction are often linked with evolution from proximal to distal parts. Depending of the size of the reservoir versus characteristic dimensions of the system, we may have to take this into account or not.
A very interesting (but uncommon) feature of this way of defining indicators by truncation is their ability to deal with nonstationarity. Structural analysis is done using (6) and taking spatial averages both on experimental and theoretical values. This point has been presented extensively in [1], so we will not enter in details here, just illustrate some typical cases.

a) Proportions varying along the vertical,

that is, the thresholds are function of z.

- Horizontal variograms depend explicitly on the level z, i.e we have one horizontal variogram per level (experimental and theoretical). Generally we fit the mean horizontal variogram averaging (6) for all possible z (see Figure 6 for an actual example)

• Vertical variograms depend explicitly on z and also on the modulus and direction (upward – downward) of h. It is only for fitting when we take the average over all z (on all the wells) that we obtain symmetrical statistics on h (see Figure 7).

b) Proportions varying along the horizontal direction

Figure 8 shows the effect of non–isotropic horizontal proportions on the indicator variograms, displaying the impact on the resulting model.

Another way to consider proportions varying in space is as additional constraints on the final model, and not only for calibration of the conditional variogram (non–stationary in this case). We believe that this way of expressing additional knowledge or constraints by proportions is well suited to the study of reservoirs. A few wells and some geological knowledge give the vertical evolution. Seismic and (or) geological knowledge can give the horizontal evolution. However we have to note that the way to build a complex model of 3D proportions can vary considerably depending on the case, knowledge, vertical resolution of seismic etc.

Conditional simulation

One of the main reasons for using truncated gaussians is that having reconstructed the gaussian values at the wells, the process for conditional simulation is well known and rigourous. Among the methods to build gaussian values at the wells, the simplest is the one based on the classical decomposition:

$$Y(x) = Y^*(x) + \sigma R(x)$$

where $Y^*(x)$ is the kriging estimate of $Y(x)$, $R(x)$ is the standardized residual and σ is the kriging standard deviation. If at location x we have facies F_i this means:

$$\frac{s_{i-1} - Y^*(x)}{\sigma} < R(x) \leq \frac{s_i - Y^*(x)}{\sigma}$$

To reconstruct these $R(x)$ we can for example use the methods presented in [3], [2] or some adequate variants.

EXTENSION OF TRUNCATED GAUSSIANS

There are many. For example we can mention substitution random functions [4]. But here, we would like to focus one particular extension: the multivariate truncated gaussian.

So far we have been truncating one gaussian random function to define facies. It is straight forward to extend this method by truncating linear combinations of p gaussian random functions. Some of them could be used to include constraints coming from other information.

The reasons for using gaussians include :

- Any linear functional of gaussian is gaussian. So the multivariate distribution involved will still be gaussian. Only the covariance matrix of these functionals has to be computed before being able to determine (at least in theory) the thresholds and indicator covariances.
- In general it is difficult to determine a consistent model for variograms and cross variograms. With gaussian random functions it suffices that the covariance (simple and cross) matrix be positive definite. Some models satisfying these constraints are known.

- At this level of generality they are heuristic algorithms for simulation.

We will now describe two of these models using two stationary zero mean gaussians Y_1 and Y_2, with unit variance, covariances ϱ_1 , ϱ_2 and cross covariance ϱ. Even in this case the freedom to choose linear functionals to truncate is large. We restrict ourselves to the simplest case.

- Truncation is done on Y_1 and Y_2. ie we define facies by conditions

$$\{x \; ; \; a < Y_1(x) \leq b \; , \; c \leq Y_2(x) \leq d\}$$

- Only one threshold s_1 for Y_1. That is, we partition our facies in two sets the first p being associated with $Y_1(x) \leq s_1$ the remaining with $Y_1(x) > s_1$

- **First example** :

Y_1 and Y_2 independent – 3 facies

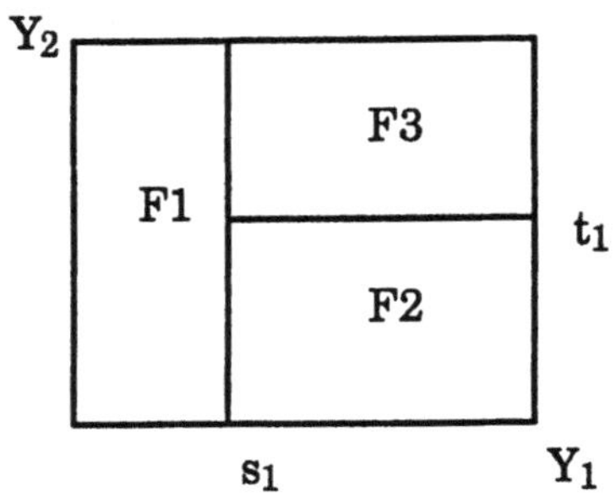

$F_1 = \{x; Y_1(x) \leq s_1\}$

$F_2 = \{x; Y_1(x) > s_1 \text{ and } Y_2(x) \leq t_1\}$

$F_3 = \{x \; ; \; Y_1(x) > s_1 \text{ and } Y_2(x) > t_1\}$

We have :

$s_1 = G^{-1}(p_1)$

$$p_2 = P(Y_1(x) > s_1 \ , \ Y_2(x) \le t_1) = P(Y_1(x) > s_1)P(Y_2(x) \le t_1) = (1 - p_1)G(t_1)$$
$$=> t_1 = G^{-1}\left(\frac{p_2}{1 - p_1}\right) = G^{-1}(\tilde{p}_2)$$
where $\tilde{p}_2 = \dfrac{p_2}{p_2 + p_3}$

Now we compute C_1 and C_2 the (non centered) covariances of F_1 and F_2.

$$C_1(h) = P(Y_1(x) \le s_1 \ , \ Y_1(x + h) \le s_1)$$

We recognize the expression obtained with only one gaussian. That is this facies will have the same anisotropy as ϱ_1 (the covariance of Y_1).

$$C_2(h) = P(Y_1(x) > s_1 \ , \ Y_1(x + h) > s_1 \ , \ Y_2(x) \le t_1 \ , \ Y_2(x + h) \le t_1)$$
$$= P(Y_1(x) > s_1 \ , \ Y_1(x + h) > s_1)P(Y_2(x) \le t_1 \ , \ Y_2(x + h) \le t_1)$$
$$= (1 - 2p_1 + C_1(h))\tilde{C}_2(h)$$

where $\tilde{C}_2(h)$ depends only on one gaussian Y_2, and will have the anisotropy of ϱ_2 (the covariance of Y_2). So C_2 will have a complex anisotropy, not only as a consequence of the Gaussian assumption (the factorization of C_2 is only a consequence of the independence between Y_1 and Y_2) but also as a consequence of consistency relations (3).
Specify the proportions, and let

$p_1 = 50\ \%$, $p_2 = 25\ \% \ => \tilde{p}_2 = 50\ \%$ so that $s_1 = 0$, $t_1 = 0$.
Using (7) we obtain

$$\sigma_1(h) = \frac{1}{2\pi}\text{Arc Sin } \varrho_1(h)$$

$$\sigma_2(h) = \left(\frac{1}{2\pi}\text{Arc Sin } \varrho_1(h) + 0.25\right)\left(\frac{1}{2\pi}\text{Arc Sin } \varrho_2(h) + 0.25\right) - (0.25)^2$$

On figure 9, we present the isocovariance curves for σ_2, where ϱ_1 and ϱ_2 are exponentials, with geometrical anisotropies (1:3) for ϱ_1 in direction 0°, (1:5) for ϱ_2 in direction 45°.

The physical interpretation of this model is an erosion of facies F_2 and F_3 by facies F_1, and the simulation can be done in two steps.

1. Simulation of facies F_2 and F_3 using Y_2 and $\tilde{C}_2$, with proportions $\tilde{p}_2$ and $(1 - \tilde{p}_2)$; information about facies F_1 is ignored.
2. Simulation of facies F_1 and its complement $\overline{F}_1$, using Y_1 with respective proportions p_1 and $(1-p_1)$ conditioned by F_1 and $\overline{F}_1(= F_2 \cup F_3)$

Finally make the erosion of facies F_2 and F_3 by facies F_1 (this means if at a pixel simulation 2 is F_1 we take F_1 ,if not we take the value of simulation 1).

- **Second example**

Y_1 and Y_2 having a cross covariance ϱ (h) – 4 facies

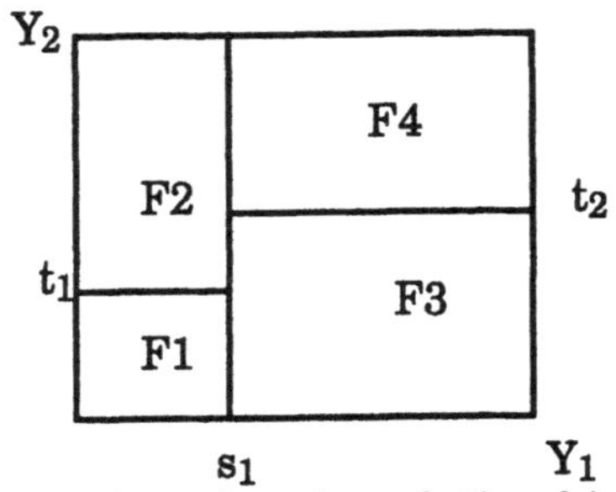

It is clear that the relationship between thresholds and proportions involves the bivariate cumulative distribution. In our case, this involves the correlation coefficient $\varrho = \varrho(o)$ between Y_1 and Y_2. Before writing these relations, we use the fact that we have only one threshold s_1 on Y_1. Using $p_1 + p_2 = G(s_1)$, we get $s_1 = G^{-1}(p_1 + p_2)$, so that s_1 is known and independent of ϱ

Now writing these relations for p_1 and p_3

$$p_1 = P(Y_1(x) \leq s_1 \ , \ Y_2(x) \leq t_1)$$

$$p_3 = P(Y_1(x) > s_1 \ , \ Y_2(x) \leq t_2)$$

We see that, for each ϱ , s_1 being known, the right hand side of p_1 is a strictly increasing function of t_1 and so can be inverted giving t_1. With the same argument we obtain t_2.. So we end up with two thresholds functions of ϱ; that is, one degree of freedom.

For the centered covariances C_{ij} we have :

$C_{ij}(h) = P(Y_1(x) \in I_i \ , \ Y_1(x+h) \in I_j \ , \ Y_2(x) \in K_i \ , \ Y_2(x+h) \in K_j)$

where I_i and I_j are $[-\infty \ , \ s_1]$, or $]s_1 \ , \ \infty[$

K_i and K_j are $[-\infty \ , \ t_1]$ $[-\infty \ , \ t_2]$, $]t_1 \ , \ \infty[$, $]t_2 \ , \ \infty[$ depending on the facies.

Consequently as $\varrho(o)$ is fixed, the indicator variograms, and cross variograms can be computed for each possible choice of $(\varrho_1(h) \ , \ \varrho_2(h) \ , \ \varrho(h))$ and compared to the experimental variogram of facies.

The idea behind this model is close to the first one. First simulate the sets $F_1 \cup F_2$ and $F_4 \cup F_3$, their covariances depending only on Y_1. Then simulate inside these sets, as shown in the figure above.

CONCLUSION

We have seen that the truncated gaussian method is a powerful tool for consistently modelling multiple facies presenting a sequential order. It is easy to use and allows us to incorporate some external constraints.

The hypothesis behind this method is clear; its main limitations have been clearly identified. It is possible (at least if enough data are available) to check if a

truncated gaussian can reasonably be used for a given case, and to estimate its parameters. Extensions proposed here make the truncated gaussian a very general tool for reservoir characterization.
However it is probably hopeless to think about a single universal method for modeling reservoirs (we believe that for a given reservoir, depending of the problem, we could have to use different methods). There is a wide domain open for new (consistent) well formulated models with enough flexibility to integrate a large amount of constraints.

REFERENCES

[1] Beucher H., Galli A., Le Loc'h G., Ravenne C. and Heresim Group (1993) "Including a regional trend in reservoir modelling using the truncated Gaussian method" Proceeding of the Fourth Geostatistics Congress Troia 1992, Kluwer Academic Publishers.

[2] Freulon X., de Fouquet C. "Conditioning a Gaussian model with inequalities" Proceeding of the Fourth Geostatistics Congress Troia 1992, Kluwer Academic Publishers.

[3] Langlais V., (1990) "Estimation sous contraintes d'inégalités" Thèse docteur en Géostatistique, ENSMP, Fontainebleau (France)

[4] Lantuejoul C. "Substitution random functions" Proceeding of the Fourth Geostatistics Congress Troia 1992, Kluwer Academic Publishers.

[5] Matheron G.,(1975) Compléments sur les modèles isofactoriels. Rapport interne N 432 – Centre de Géostatistique, ENSMP, Fontainebleau

[6] Matheron G., Beucher H., De Fouquet C., Galli A., Guerillot D., Ravenne C.,(1987) "Conditional simulation of the geometry of fluvio deltaic reservoir", SPE 16753.

[7] Matheron G., "The internal consistency of models in Geostatistics". Proc. of the third international Geostatistics congress, Sept 5–9, 1988, Avignon, France. Kluwer Academic Publishers.

[8] Ravenne C., Galli A., Mathieu Y., Montadert L., Rudkiewicz J.L. (1987) "Heterogeneities and geometry of sedimentary bodies in a fluvio–deltaic reservoir", SPE 16752,

[9] Rivoirard J."Relation between the indicators related to a regionalized variable" Proceeding of the Fourth Geostatistics Congress Troia 1992, Kluwer Academic Publishers.

ANNEXE

Basic properties of indicators

1. Case with two facies F and $\overline{F}$ $(F \cup \overline{F} = R^n)$

a) $\gamma_F = \gamma_{\overline{F}}$ (1)

because

$$1_F(x) - 1_F(x+h) = 1_{\overline{F}}(x) - 1_{\overline{F}}(x+h)$$

b) Assuming stationarity we have

$$\sigma_F(h) = \sigma_{\overline{F}}(h)$$

where σ is the centered covariance. This is just the classical relationship between variogram and covariance, keeping in mind that

$$\mathrm{Var}F = P_F(1 - P_F) = \mathrm{Var}\overline{F}$$

c) In the stationary case

$$\sigma_F(h) = \sigma_{\overline{F}}(h) = -\sigma_{F\overline{F}}(h) = -\sigma_{\overline{F}F}(h) \tag{2}$$

where $\sigma_{F\overline{F}}(h) = E\Big((1_F(x) - P_F)(1_{\overline{F}}(x+h) - P_{\overline{F}})\Big)$

and $\sigma_{\overline{F}F}(h) = E\Big((1_{\overline{F}}(x) - p_{\overline{F}})(1_F(x+h) - p_F)\Big)$

Proof.

$$\text{from} \quad E\Big((1_{F\cup\overline{F}}(x) - 1)\Big)\Big(1_{F\cup\overline{F}}(x+h) - 1)\Big) = 0$$

using $1 = P_F + P_{\overline{F}}$

and $1_{F\cup\overline{F}}(y) = 1_F(y) + 1_{\overline{F}}(y)$

we obtain $\sigma_F(h) + \sigma_{\overline{F}}(h) + \sigma_{F\overline{F}}(h) + \sigma_{\overline{F}F}(h) = 0$

from $E\Big((1_{F\cup\overline{F}}(x) - 1)(1_F(x+h) - p_F)\Big) = 0$

we get $\sigma_F(h) + \sigma_{\overline{F}F}(h) = 0$

2. Case with 3 facies A, B, C

Choosing alternatively $F = A \cup B$, $F = A \cup C$, $F = B \cup C$, we get

$$\sigma_{A \cup B}(h) = \sigma_C(h) = -\sigma_{(A \cup B)C}(h) = -\sigma_{C(A \cup B)}(h)$$

$$\sigma_{A \cup C}(h) = \sigma_B(h) = -\sigma_{(A \cup C)B}(h) = -\sigma_{B(A \cup C)}(h) \qquad \text{(I)}$$

$$\sigma_{B \cup C}(h) = \sigma_A(h) = -\sigma_{B \cup C)A}(h) = -\sigma_{A(B \cup C)}(h)$$

Using the second equalities of (I), we obtain :

$$\sigma_A(h) = -\sigma_{AB}(h) - \sigma_{AC}(h)$$

$$\sigma_B(h) = -\sigma_{BA}(h) - \sigma_{BC}(h)$$

$$\sigma_C(h) = -\sigma_{CA}(h) - \sigma_{CB}(h)$$

Using now the third equalities of (I), we have

$$\sigma_{AB}(h) - \sigma_{BA}(h) = \sigma_{BC} - \sigma_{CB}h = \sigma_{CA}(h) - \sigma_{AC}(h) \qquad (3)$$

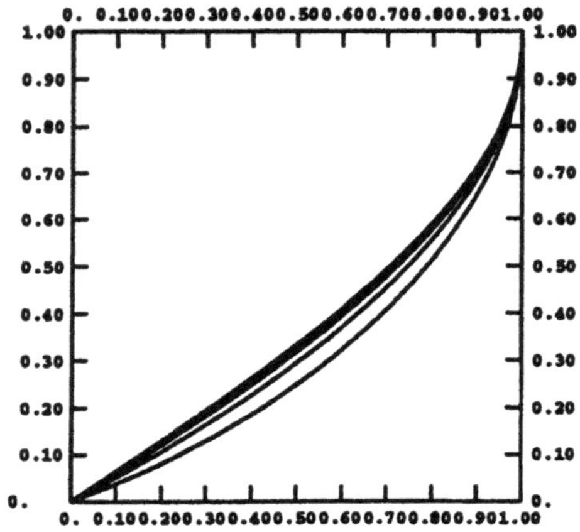

Figure 1 :
Standardized centred indicator covariance function of covariance of the Gaussian

. Case of first facies.
Proportions (from bottom to top)
10 %, 20 %, 30 %, 40 %, 50 %

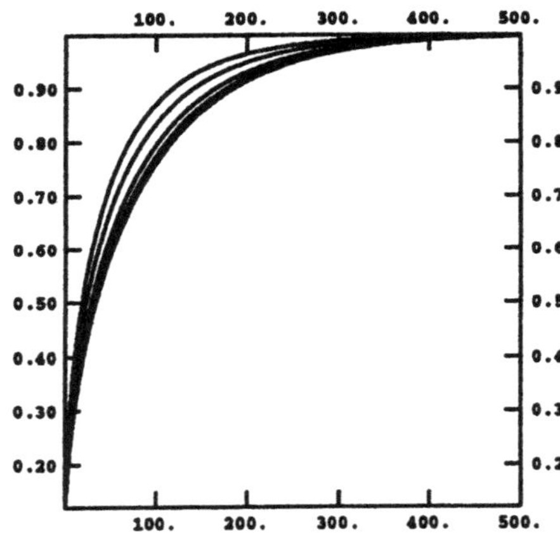

Figure 2 :
Standardized indicator variogram for the first facies.

Proportions (from top to bottom)
5 %, 10 %, 20 %, 30 %, 50 %.

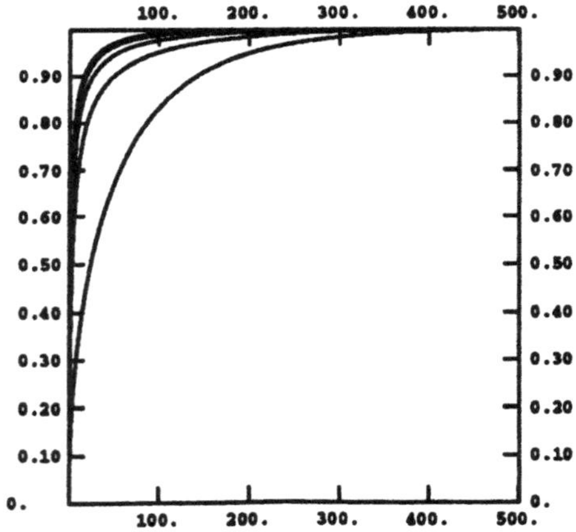

Figure 3 :
Standardized indicator variogram for the second facies with proportions 10 %,when the first facies have proportions
0 %, 10 %, 20 %, 30 %, 40 %.

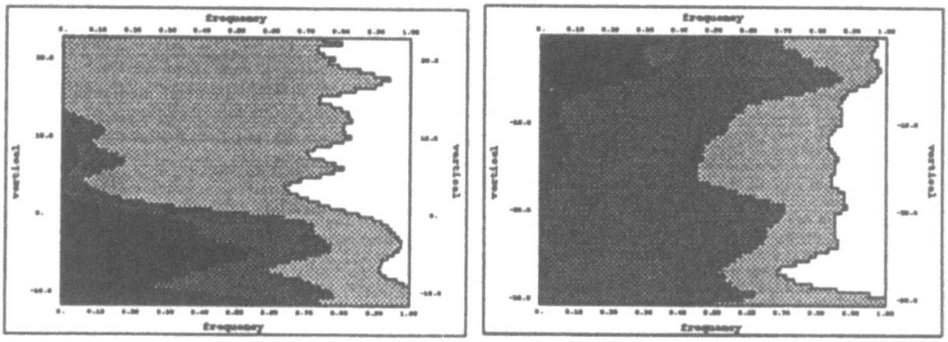

Fig 4a Fig. 4b

Figure 4 : Example of vertical proportion curves

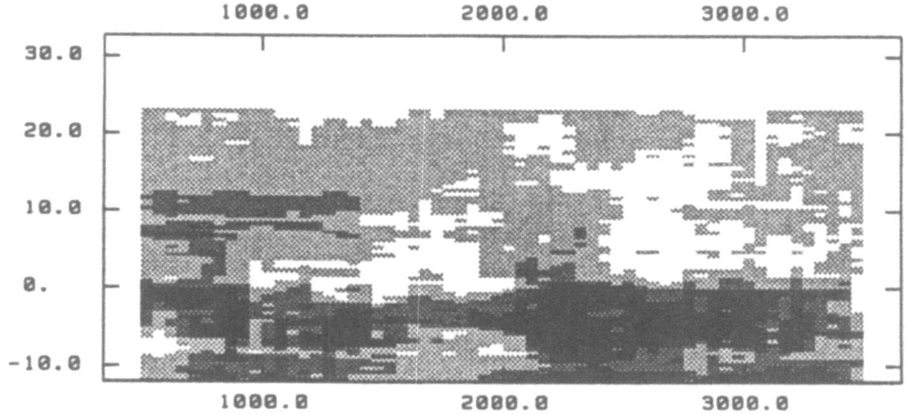

Fig. 5a : with proportions of Figure 4a

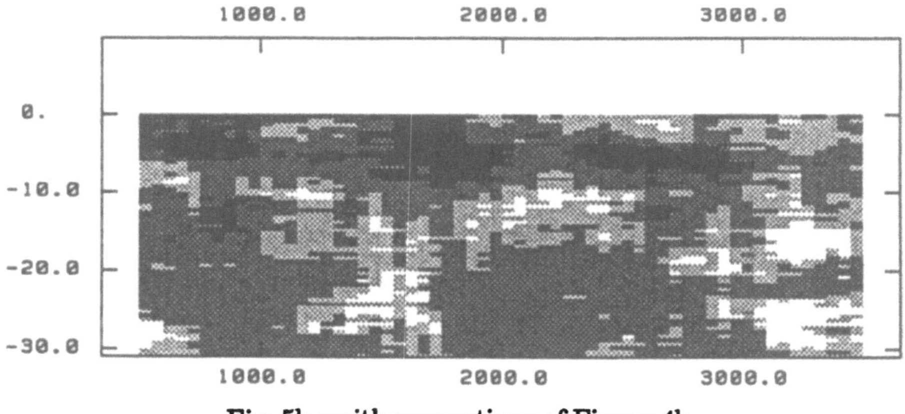

Fig. 5b : with proportions of Figure 4b

Figure 5 : Simulation with vertical nonstationarity

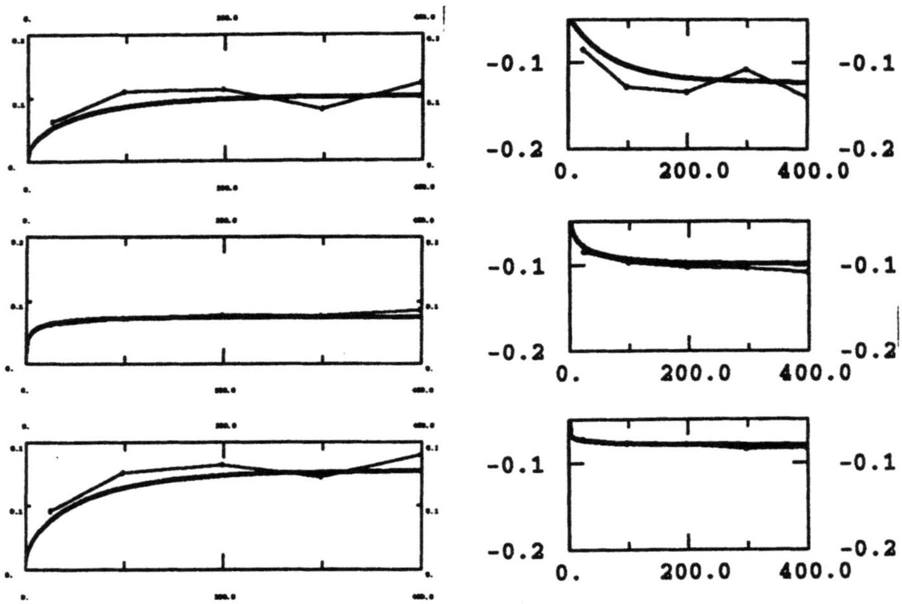

Figure 6a: simple variograms Figure 6b: cross variograms

Figure 6 : Mean horizontal variograms and their fits

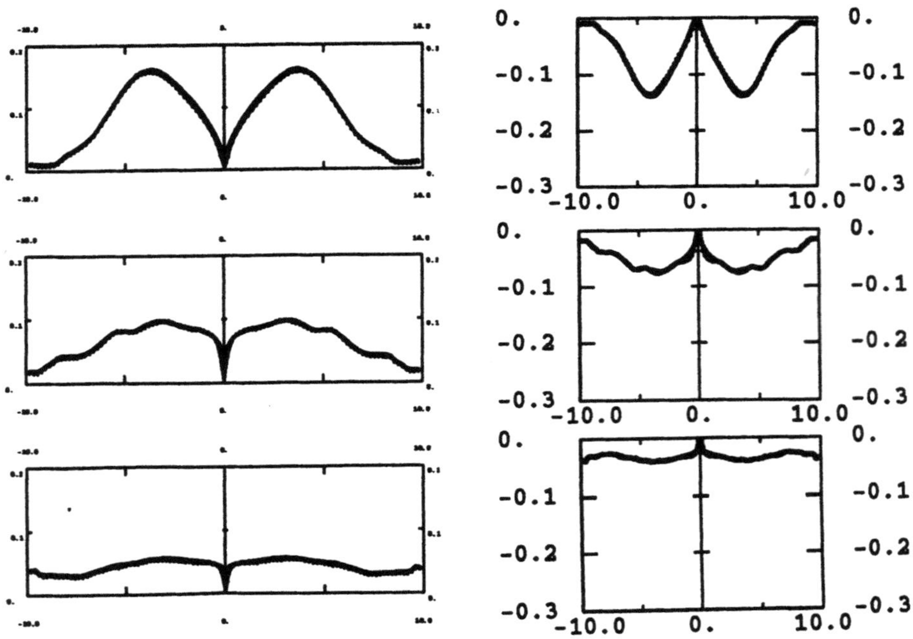

Figure 7a: simple variograms Figure 7b: cross variograms

Figure 7 : Mean vertical variograms and their fits

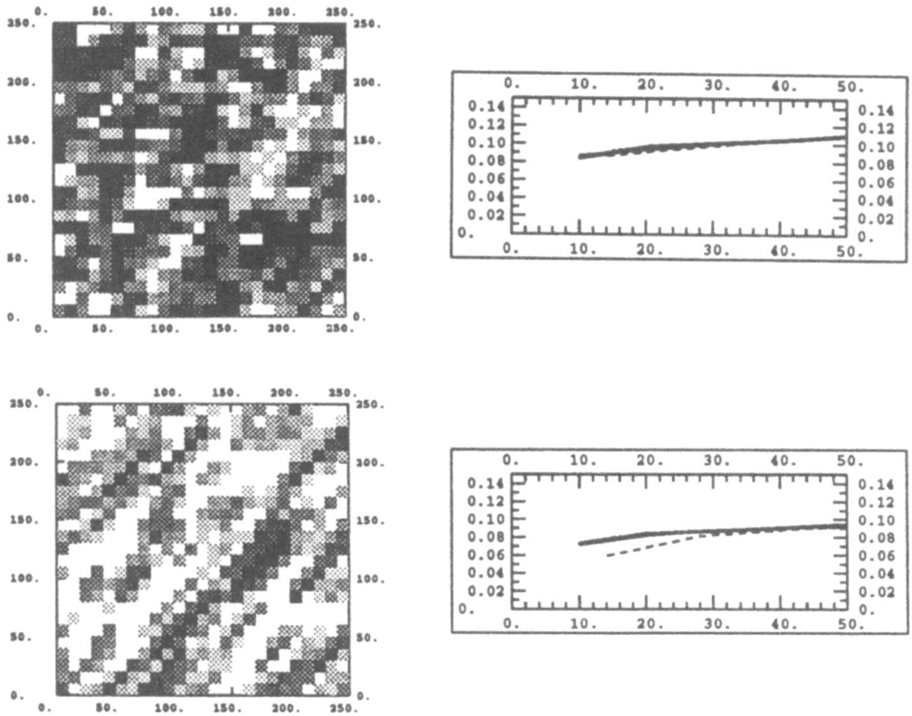

Figure 8 : Case with 3 facies. Non stationary horizontal proportions for two of them and their theoretical variograms for the four main directions.

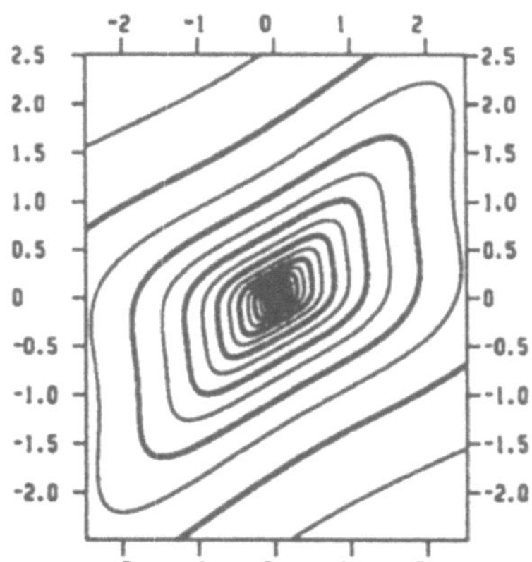

Figure 9 : Isocovariance curves for the facies corresponding to $\{x ; Y_1(x) > 0$ and $Y_2(x) < 0\}$, Y_1, Y_2 independant, their exponential covariance presenting geometrical anisotropies (1:3) for Y_1 in direction 0°, (1:5) for Y_2 in direction 45°

CONDITIONAL SIMULATION OF FACIES ARCHITECTURE IN FLUVIAL RESERVOIRS

Frode Georgsen[1], Thore Egeland[1], Ragnar Knarud[2] and Henning Omre[1,3]
[1]*Norwegian Computing Center*
P.O.Box 114, Blindern
N-0314 OSLO, Norway

[2]*Saga Petroleum a.s.*
P.O.Box 490
N-1301 SANDVIKA, Norway

[3] *University of Trondheim, NTH*
Department of Mathematics
N-7034 TRONDHEIM, Norway

1 Introduction and summary

This paper gives details of technical aspects of Georgsen and Omre [5] related to simulation algorithms based on stochastic models for fluvial reservoirs, in addition to extending the aforementioned paper in several ways.

In previous models named SISABOSA and FLUREMO, see Augedal et al. [1] and Clemetsen et al. [3], the building blocks were sandstone channels. Stacking patterns of channels, based on parameters given by the geologists, resulted in channel clustering and assumptions of channel-belt developments. The channels were homogeneous reservoir volumes in a non-reservoir mudstone matrix. Flow directions, sand body dimensions, sinuousity, sand/gross ratios etc. for both channels and channel-belts were estimated and used as input parameters to the models, based on well and field analog information.

SISABOSA and FLUREMO were developed in an exploration phase when the number of wells were limited. A strong increase in well information during field production phase caused the need for a new generation model. The requirement of this phase of development, was to bring all the specific well information into the reservoir simulation. Facies correlation between wells in a fluvial system where the well spacing is 600-1000 meters may contribute to many optional solutions. The new model is aimed at coping with these problems by linking deterministic and stochastic information.

In addition to channel-filled sands, the new model can handle crevasse-sheets and sheetsplays, together with intra-channel barriers of claydrapes, clayplugs and cement concretions. All these architectural elements can be modeled and given specific attribute values. The model is flexible and the channel-belt can be treated as

M. Armstrong and P. A. Dowd (eds.), Geostatistical Simulations, 235–250.

one volume with internal barriers or build up channel-belts by a stack of sand-filled channels where the barriers are linked to each single channel.

The general approach remains the same as in [5]. The directions of the channel-belts are represented by fibres. Geological studies provide prior knowledge about the directions. Combining this information, according to Bayesian rules, with the well-observations, gives the posterior probability distributions of the directions. The joint distribution of the fibres includes spatial interaction.

The individual channels within a channel-belt are modeled by correlated 1D Gaussian random functions. These Gaussian functions define the location, thickness and width of the channels. In addition this model handles deviated wells.

The simulation technique is based on the Ripley-Kelly algorithm for birth/death processes, Hastings algorithm with simulated annealing and a filter procedure for the 1D Gaussian random functions.

Section 2 provides a description of the stochastic model while Section 3 details the simulation algorithm. The former section is necessarily quite comprehensive and technical; a quick impression of the paper could be obtained by starting reading the part on the simulation algorithms and then reviewing Section 2 as the need arises. Case studies are provided in Section 4. Compared to similar material in [5] the present examples are based on more practical experience. Problems (arising from applying the program) and alternatives are discussed briefly in Section 5 while some concluding remarks end the paper.

2 Stochastic model for the facies architecture

The geological characteristics of the reservoir are represented by a stochastic model. In this presentation a stochastic model for the facies architecture within the reservoir is set up. This model extends the two-facies model presented in Georgsen and Omre [5].

The extended model includes totally 4 different types of facies; *background, channel sand, sheetsplay sand* and *barriers.* We will establish the model for these four types by starting with a prior model and develop it to a posterior model by introducing the well-data.

Prior model. To establish the model, some of the concepts introduced in Georgsen and Omre [5] are recalled. The simulation volume is assumed to be a rectangular reservoir-box in 3D, defined in a (x^1, x^2, x^3)-system. The channels, sheetsplays and barriers are grouped together and organized in *channel-belts.* Each of these channel-belts has a principal direction given by a line. For channel-belt number i, the principal direction line is given by l^i_s, see Figure 1.

This line is parameterized by $l^i_s : \{(x^1, x^2, x^3) : (x^1_c, \theta^i_1, \theta^i_3) + s \cdot (1, \theta^i_2, \theta^i_4)\}$, with $s \in (-\infty, \infty)$ being the argument along the line; and x^1_c being the midpoint of the reservoir-box in the x^1-direction. The x^1-direction is identified as the most

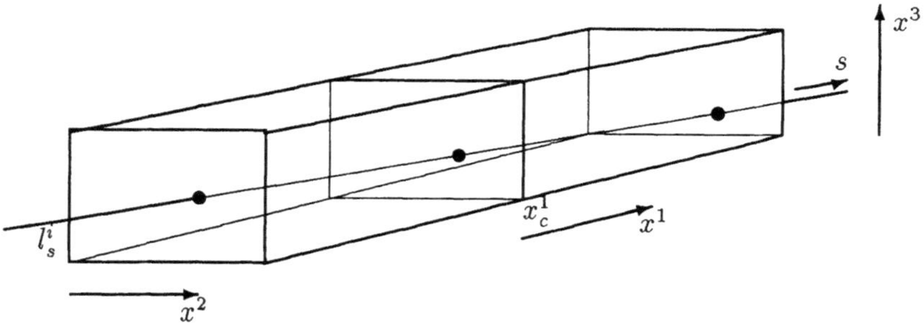

Figure 1: Principal direction line for one channel-belt, intersecting the reservoir box. The simulation volume is assumed to be a rectangular reservoir-box in 3D, defined in a (x^1, x^2, x^3)-system. Each channel-belt has a principal direction given by a line. For channel-belt number i, this is given by l_s^i.

probable direction for a channel-belt. l_s^i is fully specified by the parameter-vector $\vec{\theta^i} = (\theta_1^i, \theta_2^i, \theta_3^i, \theta_4^i)$, with θ_1^i and θ_3^i being the x^2- and x^3-coordinates respectively, of the intersection point between l_s^i and the plane $x^1 = x_c^1$; and θ_2^i and θ_4^i being the slopes of $l_{\vec{s}}^i$ projected into the (x^1, x^2)-plane and the (x^1, x^3)-plane respectively. Now consider $\vec{\theta^i}$ to be a realization of a stochastic vector $\vec{\Theta}^i$.

Consider an arbitrary channel-belt i, and let everything related to it be referenced by s along the principal direction l_s^i. Then the position of the channels in this channel-belt, at reference s, given the principal direction line is defined by a vector:

$$\{\vec{U}^i(s)|\vec{\Theta}^i = \vec{\theta^i}\} = \{U_{Hj}^i(s), U_{Vj}^i(s), U_{Wj}^i(s), U_{Tj}^i(s); j = 1, \ldots, N_U^i|\vec{\Theta}^i = \vec{\theta^i}\},$$

with N_U^i being the number of channels in the channel-belt; $U_{Hj}^i(s)$ being the horizontal sinuousity of the center line of channel j; $U_{Vj}^i(s)$ being the vertical sinuousity of the center line of channel j; $U_{Wj}^i(s)$ being the width of channel j; and $U_{Tj}^i(s)$ being the thickness of channel j, see Figure 2. From now on the reference s will be dropped when referring to the channel-belt, so $\vec{U^i}$ represents all channels in channel-belt i.

Given the principal direction line parameterized by $\vec{\theta^i}$, the vector $\vec{U^i}$ is defined as a $4N_U^i$-dimensional correlated Gaussian function, with a one-dimensional reference along l^i. The probability distribution for this function is denoted:

$$\{\vec{U}^i|\vec{\Theta}^i = \vec{\theta}^i\} \sim P_{\vec{U}^i|\vec{\Theta}^i}(\vec{u}^i|\vec{\theta}^i). \tag{1}$$

The sheetsplays are deposited on the flanks of the channels. Each sheetsplay is connected to one channel and one channel may have one or more sheetsplays associated with it. Figure 2 shows the parameterization of one sheetsplay given the channel at reference s. Let

$$\vec{V}_j^i(s) = \{V_{Hjk}^i(s), V_{Vjk}^i(s), V_{Tjk}^i(s); k = 1, \ldots, N_{Vj}^i\}$$

represent the sheetsplays associated with channel j in channel-belt i, with N_{Vj}^i being the number of sheetsplays associated with the channel; $V_{Hjk}^i(s)$ being the horizontal

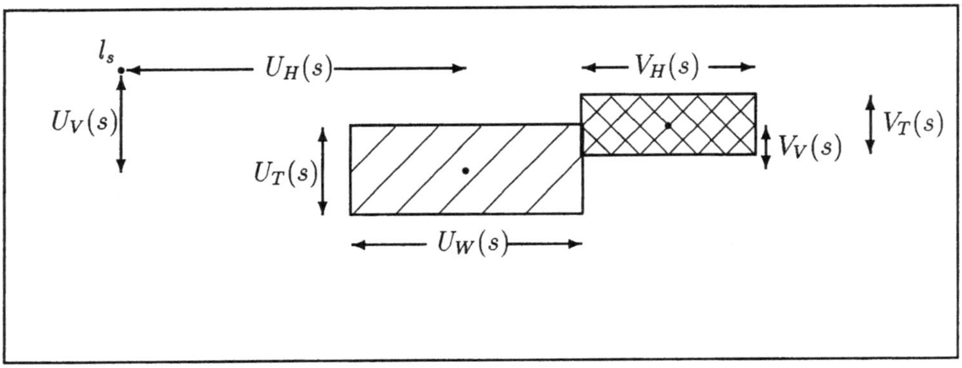

Figure 2: Cross-section of the reservoir perpendicular to the principal direction of the channel-belt, containing one channel(hatched) with one sheetsplay(crossed). The channel is given by: $U_H(s)$, the horizontal sinuousity of the center-line; $U_V(s)$, the vertical sinuousity of the center line; $U_W(s)$, the width; and $U_T(s)$, the thickness. The sheetsplay is given by: $V_H(s)$, the horizontal extension; $V_V(s)$, the vertical deviation; and $V_T(s)$, the thickness.

extension of sheetsplay k from the channel; $V^i_{Vjk}(s)$ being the vertical deviation of the center of sheetsplay k from the channel's center; and $V^i_{Tjk}(s)$ being the thickness of sheetsplay k. The probability distribution for the sheetsplays associated with channel j in channel-belt i given the channels is defined by:

$$\{\vec{V}^i_j | \vec{U}^i = \vec{u}^i\} \sim P_{\vec{V}^i_j|\vec{U}^i}(\vec{v}^i_j|\vec{u}^i). \tag{2}$$

In this distribution the V^i_{Hjk}'s are modeled as uncorrelated Gaussian functions, while the V^i_{Vjk}'s and V^i_{Tjk}'s have constant values for all s. The distribution is constrained by a strict ordering of the sheetsplays. They are not allowed to cross each other vertically.

The barriers are placed inside the channels as elliptic slices with constant thickness, relative to the center-line of the channel. This means that the shape is influenced by the sinuousity of the channels. See Figure 3 for a horizontal projection of channels, sheetsplays and barriers. The barriers are modeled as a marked point process inside the channels. Let

$$\vec{W}^i_j = \{W^i_{Cj}(k), W^i_{Lj}(k), W^i_{Wj}(k), W^i_{Tj}(k); k = 1, \dots, N^i_{Wj}\}$$

represent the barriers associated with channel j in channel-belt i, with N^i_{Wj} being the number of barriers associated with channel j; $W^i_{Cj}(k)$ being the center-point of barrier k; $W^i_{Lj}(k)$ being the length of the ellipse along the center-line of channel j; $W^i_{Wj}(k)$ being the width of the ellipse orthogonal to the center-line of channel j; and $W^i_{Tj}(k)$ being the thickness of the elliptic slice. The probability distribution for the barriers associated with channel j in channel-belt i is defined by:

$$\{\vec{W}^i_j | \vec{U}^i = \vec{u}^i\} \sim P_{\vec{W}^i_j|\vec{U}^i}(\vec{W}^i_j|\vec{u}^i). \tag{3}$$

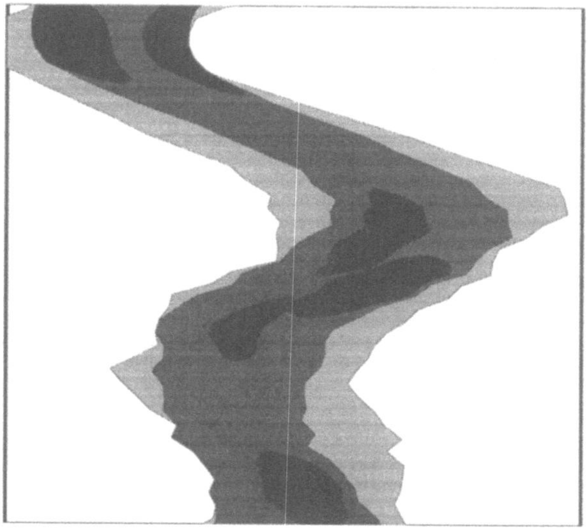

Figure 3: Horizontal projection of one channel-belt with one channel (grey), two sheetsplays (light) and six barriers (dark). Note that the elliptic shape of the barriers is transformed due to the fluctuations of the channel.

In this distribution the center-points W^i_{Cj} are distributed uniformly inside the channel, while the length, width and thickness are triangular distributed, all uncorrelated.

The ensemble of all the principal direction lines is represented in the stochastic vector $\vec{\Theta} = (\vec{\Theta}^i; i = 1, \ldots, N)$. The distribution of this vector may be defined by:

$$P_{\vec{\Theta}}(\vec{\theta}) \propto \exp\{-[\sum_{i=1}^{N} b^{\Theta}(\vec{\theta}^i) + \sum_{i=1}^{N}\sum_{j=1}^{N} c^{\Theta}(\vec{\theta}^i, \vec{\theta}^j)]\}, \tag{4}$$

with N being the number of channel-belts; $b^{\Theta}(\vec{\theta})$ being the equivalent of a prior Gaussian marginal distribution and $c^{\Theta}(\cdot,\cdot)$ being the pairwise interaction between the principal directions of the channel-belts. The form of the interaction function $c^{\Theta}(\cdot,\cdot)$ can be chosen according to the spatial pattern the fibres are supposed to show. When the fibres represent channel-belts in a fluvial reservoir, geological knowledge suggests to use a function where the fibres repulse each other, see Clemetsen et al. [3].

The number of channels inside the channel-belt, N^i_U should be considered as a stochastic variable, so the distribution in expression (1) is given N^i_U. Similar comments apply for the number of sheetsplays in the channels N^i_{Vj} (2), the number of barriers in the channel N^i_{Wj} (3) and the number of channel-belts in the reservoir N (4).

The collection of all channels, sheetsplays and barriers is denoted

$$\{\vec{U}, \vec{V}, \vec{W}\} = \{\vec{U}^i, \vec{V}^i_j, \vec{W}^i_j; j = 1, \ldots, N^i_U, i = 1, \ldots, N\},$$

and the following probability distribution is obtained:

$$\begin{aligned}\{\vec{U},\vec{V},\vec{W}\} &\sim P_{\vec{U},\vec{V},\vec{W}}\{\vec{u},\vec{v},\vec{w}\}\\ &= \{\int_\Omega P_{\vec{\Theta}}(\vec{\theta})\prod_{i=1}^{N} P_{\vec{U}^i|\vec{\Theta}^i}(\vec{u}^i|\vec{\theta}^i)d\vec{\theta}\}\{\prod_{i=1}^{N}\prod_{j=1}^{N_U^i} P_{\vec{V}_j^i|\vec{U}^i}(\vec{v}_j^i|\vec{u}^i)\cdot P_{\vec{W}_j^i|\vec{U}^i}(\vec{w}_j^i|\vec{u}_j^i)\}. \end{aligned} \quad (5)$$

This expression comes from the assumption that the channel-belts only influence each other through their principal directions and that the sheetsplays and barriers in different channels are independent of each other. This expression is the conditional distribution for channels and sheetsplays given all the numbers of the facies bodies. Including the distribution for these and the need for writing (5) on a more compact form, gives the reformulation:

$$P_{\vec{B}}(\vec{b}) \propto P_N(n)\cdot \exp\{-[\sum_{i=1}^{n} Z_1(\vec{b}^i) + \sum_{i=1}^{n}\sum_{j=1}^{n} Z_2(\vec{b}^i,\vec{b}^j)]\}, \quad (6)$$

where $\vec{B} = \{\vec{B}^i; i = 1,\dots,N\} = \{\vec{U},\vec{V},\vec{W}\}$ is the ensemble of all channel-belts in the reservoir and $P_N(n)$ is the marginal distribution of the number of channel-belts.

Posterior model. Observations from wells give additional information about the parameters in the model and about the exact location of channels. Denote the number of wells in the reservoir by M. The observations appear in wells defined by piecewise linear paths through the reservoir. In each well the intersections with the different facies-bodies are identified, i.e. center-location x_k^j, facies-type f_k^j and thickness t_k^j for facies-body j, in well k.

Consider one channel-belt with principal direction l^i. Define M points $\{x_k; k = 1,\dots,M\}$ where the M wells intersect the plane going through l^i, with the unit normal-vector with maximal vertical component. In a new coordinate system in this plane we obtain $\{s_k^i = (s_k^{1,i}, s_k^{2,i}); k = 1,\dots,M\}$, with $s_k^{1,i}$ being the reference along l^i, and $s_k^{2,i}$ being the reference along the normal to l^i in this plane. For each well k and facies-body j, identify,

$$\tau_j^i(s_k^i) = \begin{cases} 0 & \text{if well } k \text{ does not penetrate facies-body } j. \\ t>0 & \text{if well } k \text{ penetrates facies-body } j \text{, with thickness } t = t_k^{i,j}. \end{cases}$$

$$\gamma_j^i(s_k^i) = \begin{cases} \infty & \text{if well } k \text{ does not penetrate facies-body } j. \\ c & \text{if well } k \text{ penetrates facies-body } j \text{, with } c = x_j^{3,i} - \theta_3^i - s_k^{1,i}\cdot\theta_4^i, \\ & \text{being the depth of the channel observation relative to the} \\ & \text{principal direction of the channel-belt.} \end{cases}$$

This defines the set of available observations, which has to be pre-identified for each i, j and k. Let N_F^i denote the total number of facies bodies in channel-belt i:

$$\vec{o} : \{\tau_j^i(s_k^i), \gamma_j^i(s_k^i); k = 1,\dots,M; j = 1,\dots,N_F^i; i = 1,\dots,N\}.$$

Let P^* define the posterior analog to the probability distribution P. Then the posterior version of the probability distribution in expression (5) is defined by:

$$\{\vec{U},\vec{V},\vec{Y}|\vec{O}\} \sim P^*_{\vec{U},\vec{V},\vec{Y}}\{\vec{u},\vec{v},\vec{y}\}$$
$$= \{\int_\Omega P^*_{\vec{\Theta}}(\vec{\theta})\prod_{i=1}^{N} P^*_{\vec{U}^i|\vec{\Theta}^i}(\vec{u}^i|\vec{\theta}^i)d\vec{\theta}\}\{\prod_{i=1}^{N}\prod_{j=1}^{N_U^i} P^*_{\vec{V}_j^i|\vec{U}^i}(\vec{v}_j^i|\vec{u}^i)\cdot P^*_{\vec{W}_j^i|\vec{U}^i}(\vec{w}_j^i|\vec{u}_j^i)\}.$$

where, by omitting the indexes of the elements in $\{\vec{U},\vec{V},\vec{W}\}$:

- $P^*_{\vec{\Theta}}(\vec{\theta})$ is the posterior analog to the joint distribution for the principal direction lines given in expression (4). The posterior analog to the term $b^\Theta(\vec{\theta}^i)$ can be found analytically by making three assumptions; (i) the conditioning is only performed on the wells penetrating the facies-bodies, i.e. on the set $\{s_k^i : \tau_j^i(s_k^i) > 0; k = 1,\ldots,M; j = 1,\ldots,N_F^i\}$; (ii) the wells hit the channels at their center-line also if the observed facies is a sheetsplay or barrier; and (iii) the vector $\vec{U}^i$ is a Gaussian random function in the reservoir x-system of axis, which can be shown not to be exactly true. None of the assumptions above are considered to be crucially violated in the modeling of fluvial reservoirs. These assumptions justify the application of the methodology of Bayesian kriging, see Omre and Halvorsen [12]. The posterior analog to the term $c^\Theta(\cdot,\cdot)$ is defined to be equal to the prior. This entails that the observations do not influence the estimation of the spatial interaction relations for the principal directions.

- $P^*_{\vec{U}^i|\vec{\Theta}^i}(\vec{U}^i|\vec{\theta}^i)$ is the probability distribution of a $4N_U^i$-dimensional Gaussian random function. For each element in $\vec{o}$ one has, either; (i) the well penetrates the channel: $\tau_j^i(s_k^i) > 0$, which gives an inequality constraint, $s_k^{2,i} \in \{U_H(s_k^{1,i}) \pm \frac{U_W(s_k^{1,i})}{2}\}$, and the equality constraints $\tau_j^i(s_k^i) = U_T(s_k^{1,i})$ and $\gamma_j^i(s_k^i) = U_V(s_k^{1,i})$; or (ii) the well does not penetrate the channel: $\tau_j^i(s_k^i) = 0$, which gives the inequality, $s_k^{2,i} \notin \{U_H(s_k^{1,i}) \pm \frac{U_W(s_k^{1,i})}{2}\}$.

- $P^*_{\vec{V}_j^i|\vec{U}^i}(\vec{v}_j^i|\vec{u}^i)$ is the posterior distribution of the sheetsplays given the channels. All sheetsplays are independent of the other. Each sheetsplay is identified to be either at the negative ($I = -1$) or the positive ($I = 1$) side of the channel in the (s^{1i}, s^{2i})-system. For each element in $\vec{o}$ one has, either: (i) the well penetrates the sheetsplay: $s_k^{2,i} \leq V_H(s_k^{1,i})$ if $I = 1$, or $s_k^{2,i} \geq V_H(s_k^{1,i})$ if $I = -1$, $\gamma_j^i(s_k^i) = V_V(s_k^{1,i})$, $\tau_j^i(s_k^i) = V_T(s_k^{1j})$; or (ii) the well does not penetrate the sheetsplay: $s_k^{2i} > V_H(s_k^{1j})$ if $I = 1$, $s_k^{2i} < V_H(s_k^{1j})$ if $I = -1$. The horizontal extension of the sheetsplay is a Gaussian random function truncated at zero from below ($I = -1$) or above ($I = 1$). The vertical deviation and thickness are linearly interpolated between the conditioning values.

- $P^*_{\vec{W}^i_j|\vec{U}^i}(\vec{w}^i_j|\vec{u}^i)$ is the posterior distribution of the barriers given the channels. For each element in $\vec{o}$ one has, either; (i) the well penetrates the barrier: $W_T(k) = \tau^i_j(s^i_k)$ and the size and center-point of the ellipse is sampled so that the point (s^{1i}_k, s^{2i}_k) falls inside and the ellipse does not coincide with other wells; (ii) the well does not penetrate the barrier: size and center-point sampled from the prior distribution, constrained by the position of all wells.

This can be formulated on a compact form analogously to expression (6) as:

$$P^*_{\vec{B}}(\vec{b}) \propto P^*_N(n) \cdot \exp\{-[\sum_{i=1}^{n} Z^*_1(\vec{b}^i) + \sum_{i=1}^{n}\sum_{i=1}^{n} Z^*_2(\vec{b}^i, \vec{b}^j)]\}, \tag{7}$$

The analytical evaluation of the probability density function (7) is not possible, but simulation techniques can be used to sample from this distribution, as described in the next section.

3 Simulation algorithm

In order to generate realizations of $\vec{B}$ from expression (7), it is necessary to find a reliable stochastic simulation algorithm. The algorithm should ensure that; (i) every realization reflects the stochastic properties of the model; (ii) every realization reproduces all observations in the reservoir; and (iii) all realizations are stochastically independent of each other. In addition, performance time is an important subject, due to the size of the reservoir.

The convergence criterion for the simulation algorithm is that the simulated sand/gross ratio (sg) is within some user-specified tolerance limits ($sg^* \pm \epsilon_{sg}$). The sand/gross ratio is here defined to be the proportion of the total simulation volume occupied by channel sand.

The implementation of the model is organized in two separate modules. The first simulating channels and sheetsplays until the specified sand/gross ratio is achieved. The second simulating barriers inside the channels until a user-specified proportion of the channels are occupied by barriers.

3.1 Simulation of channels and sheetsplays

The simulation in this module is organized in three stages:

a. Simulating channel-belts independently.

b. Ripley-Kelly algorithm.

c. Hastings algorithm with simulated annealing.

At each stage a three step procedure is followed. Firstly the principal direction lines of the channel-belts are drawn (from $P^*_{\vec{\Theta}}(\cdot)$); secondly the channels given the principal direction lines are drawn (from $P^*_{\vec{U}^i|\vec{\Theta}^i}(\cdot)$) and thirdly the sheetsplays are drawn given the channels (from $P^*_{\vec{V}^i_j|\vec{U}^i}(\cdot)$).

a. Simulating channel-belts independently

This part does not account for interaction between channels-belts. Channel-belts are placed one at a time in the reservoir until sg exceeds sg^*. Channel-belt i is drawn from a probability distribution proportional to $\exp\{-Z_1^*(\vec{b}^i)\}$ i.e. independently of the other channel-belts.

b. Ripley-Kelly algorithm

This part accounts for the interaction pattern given by the function $Z_2^*(\cdot,\cdot)$ in expression (7). The number of channel-belts (N) remains constant through this part. The initial state of this part is the outcome of part a. The algorithm is due to Kelly and Ripley [9] and is briefly described as follows:

- While ($i <$ specified number of iterations ($c \cdot N$) where c is user specified) do

  Draw a number k uniformly from $\{1, \ldots, N\}$.

  Remove $\vec{b}^k$ from the ensemble of lines.

  * Generate a new channel-belt $\vec{b}^k_*$ without interaction with other lines.

  Accept $\vec{b}^k_*$ with probability $p_* \propto \exp\{-[Z_1^*(\vec{b}^k_*) + \sum_{i \neq k} Z_2^*(\vec{b}^k_*, \vec{b}^i)]\}$,

  If not accept go to $*$.

- End of iteration.

c. Hastings algorithm with simulated annealing

In this part of the simulation algorithm flexibility in the number of channel-belts is introduced, since this number is a stochastic variable. The algorithm used is due to Hastings [6]. It is a generalization of the Metropolis-algorithm given in Metropolis et al. [11]. The algorithm is combined with simulated annealing, a technique known from optimization, see Geman and Geman [4] or Hegstad et al. [7]. The idea is to introduce a "temperature"-term which is gradually lowered in a controlled manner. Theoretically the decrease should be logarithmic to ensure that the global optimum is achieved, but due to limited computer resources the scheme $T(t) = T_0\alpha^t$ is used. Here t is the index of the iteration; T_0 is the start-temperature; and α is the temperature-rate. The idea is that part a. and b. above has created a good starting point for the simulated annealing so that the error introduced by exponential decrease of the temperature is acceptable.

Let Ω denote the space of all possible constellations of channel-belts. In order to sample from the joint posterior distribution for the channel-belts $P^*_{\vec{B}}(\vec{b})$, a Markov chain on Ω, which has $P^*_{\vec{B}}(\vec{b})$ as its stationary distribution, is constructed. According to Hastings the following two-step procedure is used:

(i) Let $N(k)$ denote the number of channel-belts in state $k \in \Omega$ and sg_k the sand/gross ratio of state k. An irreducible Markov chain on Ω, with transition matrix $Q = \{q_{ij}\};\ i, j \in \Omega$ is defined as follows:

$$q_{ij} \propto \begin{cases} \frac{1}{N(i)} \exp\{-\frac{sg^*-sg_i}{T(t)}\} & \text{if } N(j) = N(i) - 1 \\ \frac{1}{N(i)} \exp\{-Z_1^*(\vec{b}^{I(j)})\} & \text{if } N(j) = N(i) \\ \exp\{\frac{sg^*-sg_i}{T(t)}\} \exp\{-Z_1^*(\vec{b}^{I(j)})\} & \text{if } N(j) = N(i) + 1 \\ 0 & \text{otherwise.} \end{cases}$$

where $\frac{1}{N(i)}$ represents the probability of drawing one of $N(i)$ channel-belts at random and removing it; the terms $\exp\{-\frac{sg^*-sg_i}{T(t)}\}$ and $\exp\{\frac{sg^*-sg_i}{T(t)}\}$ represents a penalty for sand/gross ratios deviating from the target; $I(j)$ represents the channel-belt drawn to be added to the ensemble; and $\exp\{-Z_1^*(\vec{b}^{I(j)})\}$ represents the marginal probability for this new channel-belt.

From this Markov chain one potential new state is drawn.

(ii) The probability of accepting the new state j is given by:

$$\alpha_{ij} = \min(1, \frac{\pi_j q_{ji}}{\pi_i q_{ij}}),$$

where the introduction of simulated annealing gives a revised distribution:

$$\pi_k = P^{*A}_{\vec{B}}(\vec{b}) \propto \exp\{-[\sum_{i=1}^{N} Z_1^*(\vec{b}^i) + \sum_{i=1}^{N}\sum_{j=1}^{N} Z_2^*(\vec{b}^i, \vec{b}^j) + \frac{(sg^* - sg_k)^2}{T(t)}]\},$$

where the term $(sg^* - sg_k)^2/T(t)$ is considered a penalty function. Note that all terms involving $Z_1^*(\cdot)$ cancels in α_{ij}, so the probability is proportional to a number depending only on the interaction and the sand/gross ratio.

The procedures (i)-(ii) are performed iteratively until sg is within the tolerance limits $(sg^* \pm \epsilon_{sg})$.

Simulation of one channel-belt The simulation of sheetsplays and channels in one channel-belt is performed on a discrete one-dimensional grid along the principal direction line for the channel-belt. The simulation of the facies-bodies consists of three separate parts:

(i) Unconditional simulation of uncorrelated Gaussian functions with a specified spatial covariance-function.

(ii) Linear combinations of the uncorrelated functions to match the correlation structure between the different functions constituting the channels in the channel-belt.

(iii) Using interpolation techniques to condition the functions on the well-observations.

For (i) a filter method with a Gaussian covariance-function is used, see Omre et al. [13] and Matérn [10]. This covariance-function is defined as:

$$C(r) = \sigma^2 \exp\{-3(\frac{r}{\alpha})^2\}$$

where the argument r is the distance between the projection of two points on the principal direction line.

For each channel the M wells creates $4 \cdot M$ conditioning values. Some of these are given by equalities and some by inequalities. By performing rejection sampling from a conditional multi-Gaussian distribution, conditioned on the equality constraints, a total of $4 \cdot M$ equality conditioning values are obtained. Then the Gaussian functions for the channels are simulated conditioned on these.

The simulation of the conditional Gaussian functions in (iii) follows the traditional lines given in Journel and Huijbregts [8]. The sheetsplays are simulated conditioned on the channel. First the number of sheetsplays is drawn from the posterior distribution, and then what side of the channel each of them are located. The Gaussian function giving the horizontal extension is simulated conditioned on both penetrating and non-penetrating wells along the same lines as for channels.

3.2 Simulation of barriers

The simulation of barriers is performed in two parts according to the two sections below. First the barriers are placed in the channels according to the well-observations, then additional barriers are placed until a specified proportion of the channels is filled up with barriers. The simulation is performed on each channel independently of all the others.

3.2.1 Conditioning on wells

The simulation of observed barriers proceeds in three steps, which are performed for all observed barriers one at a time.

(i) Draw length and width of the ellipse independently of the observation. Set the thickness equal to the thickness of the observation.

(ii) An ellipse is generated with center-point at the midpoint of the observation. Since the well does not necessarily hit the barrier at its center-point, the center-point for the barrier is drawn at random inside this ellipse.

(iii) The sampled barriers needs to be consistent with other wells. If another well hits the barrier it is resampled according to (i)–(ii).

3.2.2 Unconditional sampling of barriers

Barriers are sampled one at a time until the pre-described proportion of channels is obtained. The steps (i) and (ii) above are slightly modified:

(i) Both length, width and thickness are drawn from their distributions.

(ii) The center-point is drawn uniformly distributed along the channel.

4 Examples

The intention of this section is to illustrate the theory presented in the previous sections by examples. Two kinds of examples are shown. One synthetic with a few number of facies-bodies and one real from a fluvial north sea field. The aim of these examples is to emphasize the following features; location and behaviour of facies-bodies; relation between different types of facies and conditioning on well-observations.

Figure 4 shows the horizontal projections of two realizations from a model with two channel-belts; one channel in each; 2-4 sheetsplays for each channel and a certain proportion of the channels occupied by barriers. There is a total of 14 wells in the reservoir. Some are penetrating both channels and barriers; some are penetrating channels alone; some are penetrating sheetsplays; and one is not penetrating any facies at all.

Table 1 shows some of the parameters used in this example. Recall from Section 2 that U_W is the width of the channel; U_H is the horizontal sinuousity of the channel; V_H is the horizontal extension of the sheetsplay; W_L is the length of the barrier; and W_W is the width of the barrier. $E(\cdot)$ means the expected value of a variable, while $sd(\cdot)$ is the standard deviation. $tri(\cdot)$ gives the min, mode and max of a triangular distribution, while $range(\cdot)$ is the practical range of a Gaussian covariance function. The parameter $prop(barrier)$ gives the total proportion of channels which is occupied by barriers. In the plots this may seem larger than 10%, but that is due to the smaller thickness for barriers than for channels. This is not seen in the horizontal projection.

The two parallel series of plots are from the same model, the only difference between them is the start-seed for the random number generator. Note the larger fluctuations of the sheetsplays compared to the channels, due to the shorter range for the horizontal extension.

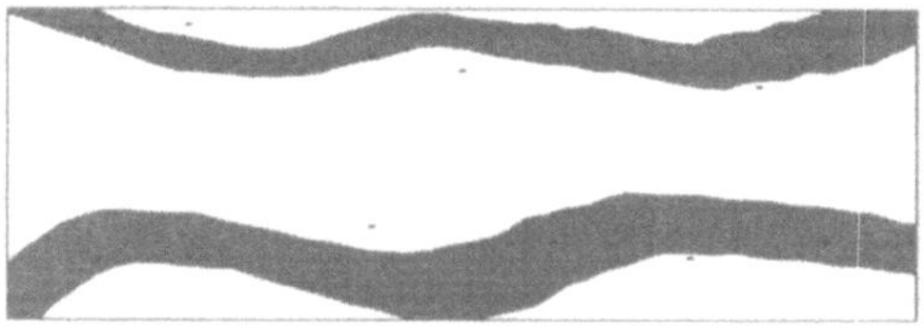

(a) Only channels

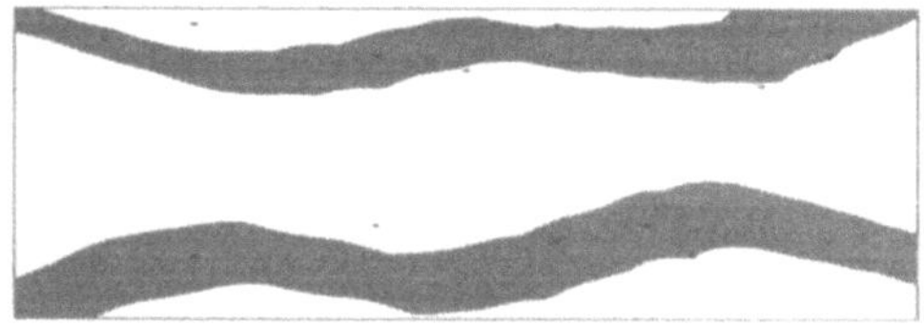

(d) Only channels

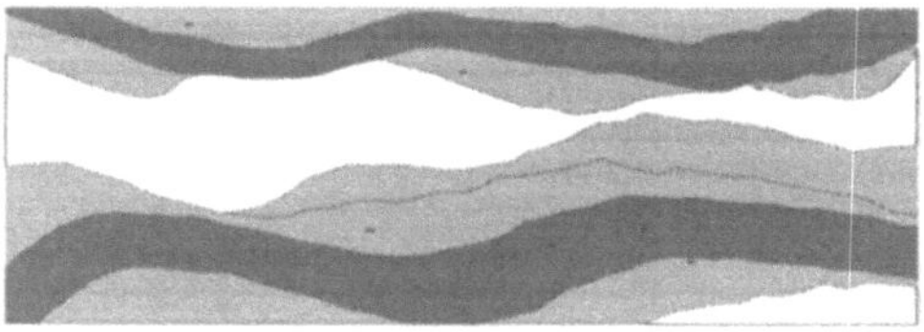

(b) channels and sheetsplays

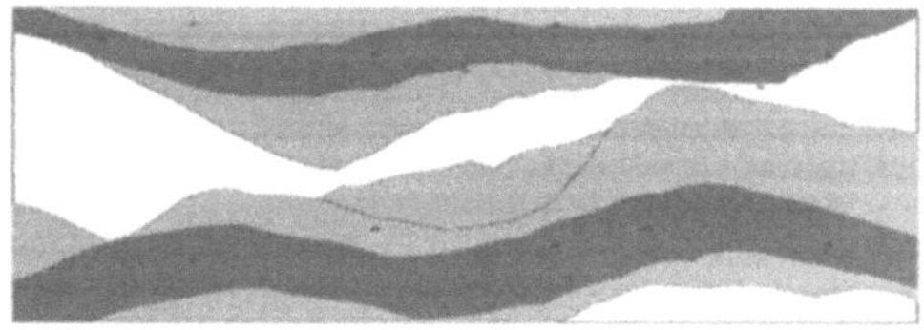

(e) channels and sheetsplays

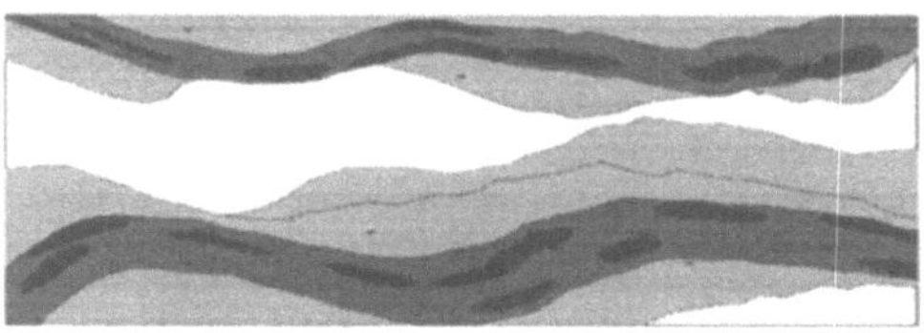

(c) channels, sheetsplays and barriers

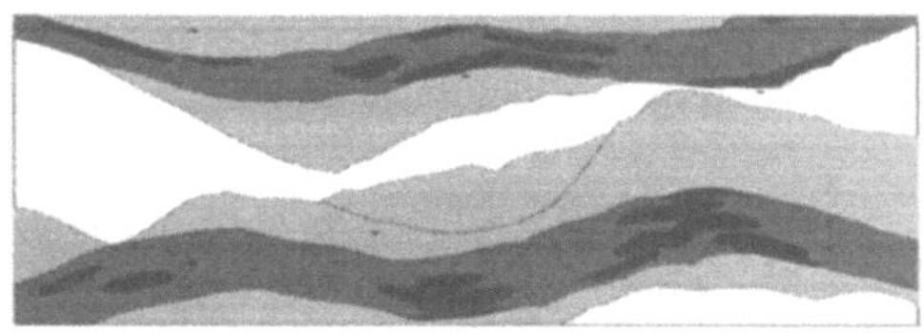

(f) channels, sheetsplays and barriers

Figure 4: Horizontal projections of two different realizations from a model with channels, sheetsplays and barriers. The parameters are given in Table 1. The wells are marked by dark dots.

All units are in meters			
channels		Sheetsplays and barriers	
Res. length	3000	$E(V_H)$	100
Res. width	1000	$sd(V_H)$	100
$E(U_W)$	200	$range(V_H)$	900
$sd(U_W)$	30	$tri(W_L)$	(300,600,900)
$sd(U_H)$	100	$tri(W_W)$	(60,110,120)
$range(U_W)$	1500	$prop(barrier)$	0.1
$range(U_H)$	1500		

Table 1: Prior parameters for the model shown in Figure 4.

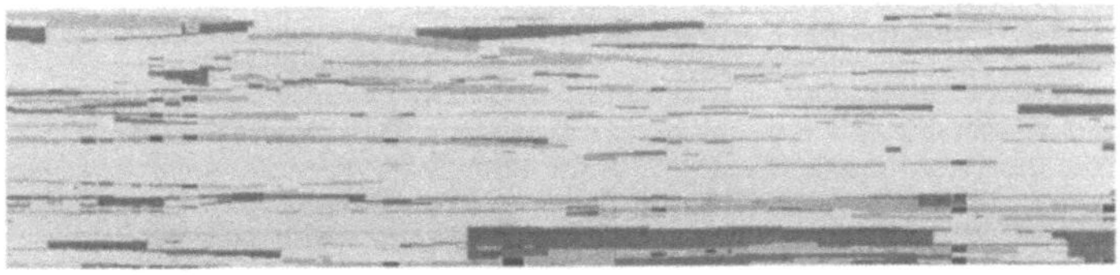

(a) Cross-section parallel to main direction of channel-belts (length = 5500m, thickness = 400m).

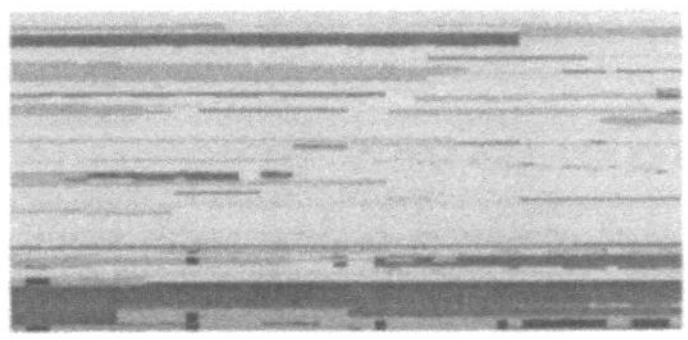

(b) Cross-section perpendicular to main direction of channel-belts (length = 1900m, thickness = 400m).

Figure 5: Cross-sections from a shale-prone part of a simulated north sea reservoir. From light to dark: Background, channels, sheet-splays and barriers.

Figure 5 shows two cross-sections of a simulated realization from a shale-prone part of a north sea reservoir with 22 vertical layers. The two plots show cross-sections along and normal to the main direction of the channel-belts. The length of the reservoir in the first plot is 5500 meters, while the length in the second is 1900. Note that all data are transformed to a rectangular box, so the variations in depth and thickness of the simulation reservoir is not accounted for in the figure. The four levels of grey-tones is from lightest to darkest; background, channels, sheet-splays and barriers. The thickness of the simulated reservoir is about 400 meters, so note that the scales in the plots are not 1:1 between the horizontal and vertical direction. The discontinuous behaviour of the channels along the main direction is due to the sinuousity, which means that the same channels may appear at repeated places in the plot. The expected width of the channels is 2 km which is about the same as the width of the cross-section in (b). That explains the continuity in this plot.

5 Problems and alternatives

Problem: Well projection. Some numerical problems arise when conditioning wells are positioned towards the principal direction of the channel-belt, that is when their projected values on the principal direction line are close even if the spatial distance between the wells is not small. This is due to the one-dimensional reference of the Gaussian functions along this line, and the smoothness of the Gaussian covariance-functions.

Problem: Well correlations. The correlation of facies-bodies in different wells

needs to be specified as input to the model. In many cases this is difficult or impossible and an automatic generation of likely well-combination would have been preferred.
Problem: Conditioning. The conditioning is not performed theoretically correct, since the sampling distribution in wells is not identical to the marginal distribution for the marks, see Chessa [2]. The larger bodies are over-represented in the sampling distribution. The errors being made are not expected to be crucial however. This is a familiar problem known from stochastic process theory. The solution can probably be found by simulating from the sampling distribution when generating the units being penetrated by wells.
Alternative: Directions. There is an obvious alternative to modeling directions by utilizing results from the statistical branch known as directional statistics. This theory is well-known in 2D but a magnitude more difficult in 3D according to Chapter 10 of Upton and Fingleton [14]. Directions for the fibres could be modeled stochastically by defining appropriate probability distributions for angles directly rather than through the parameterizations of directions. However, as indicated this may well be difficult. Besides, the present approach functions reasonably well in addition to being intuitive and intelligible.
Alternative: Time series. Once the fibres are given, multiple time-series may be used to model the geometry and sinuousity of channels and sheetsplays. Time corresponds to distance measured from some origin and is measured discretely as in the previously documented model. The main difference compared to the model presented here is that the time series is defined by an equation rather than Gaussian functions. The main advantages being that problems of non-positive-definite covariance matrices are likely to disappear and that the technical simulation will be simplified. In general, equations are easier to relate to than say densities. However, it is not entirely obvious how the equation should be set up and conditioning may pose problems. The approach would be interesting to try out if time and money were a minor concern.

Acknowledgments

The research has benefited by advice from Lars Holden of NR in addition to financial support from Saga Petroleum a.s. and the Joule–II project. Øivind Skare's programming and help with the examples is gratefully acknowledged.

References

[1] H. O. Augedal, K. O. Stanley, and H. Omre. Sisabosa, a program for stochastic modelling and evaluation of reservoir geology. In *Proceedings from Conference on Reservoir Description and Simulation with Emphasis on EOR, Oslo*, Oslo, Norway, September 1986. IFE.

[2] A. G. Chessa. On the object based method for simulating sandstone deposits. In M. A. Christie et al., editor, *Proceedings from 3rd European Conference on the Mathematics of Oil Recovery*, pages 67–78, Delft, Nederland, 1992.

[3] R. Clemetsen, A. R. Hurst, R. Knarud, and H. Omre. A computer program for evaluation of fluvial resèrvoirs. In A. T. Buller, E. Berg, O. Hjelmeland, J. Kleppe, O. Torsæter, and J. O. Aasen, editors, *North Sea Oil and Gas Reservoir – II*, pages 373–385. Graham & Trotman, May 1989.

[4] S. Geman and D. Geman. Stochastic relaxation, gibbs distributions, and the bayesian restoration of images. *IEEE Transactions on Pattern Analysis and Machine Intelligence*, Pami-6, No. 6, November 1984.

[5] F. Georgsen and H. Omre. Combining fibre processes and gaussian random functions for modelling fluvial reservoirs. In A. Soares, editor, *Geostatistics Tróia '92*, pages 425–439, Troia, Portugal, 1992.

[6] W. K. Hastings. Monte carlo sampling methods using markov chains and their applications. *Biometrika*, 57,1:97–109, 1970.

[7] B. K. Hegstad, H. Omre, H. Tjelmeland, and K. Tyler. Stochastic simulation and condition by annealing in reservoir description. In *Workshop on Geostatistical Simulations*, Fontainebleau, France, May, 1993.

[8] A. G. Journel and Ch. J. Huijbregts. *Mining Geostatistics.* Academic Press, London, 1978.

[9] F. P. Kelly and B. D. Ripley. A note on strauss's model for clustering. *Biometrika*, 63,2:357–360, 1976.

[10] B. Matérn. Spatial variation. Technical Report 5,49, Swedish Forest Research Institute, 1960.

[11] N. Metropolis, A.W. Rosenbluth, M.N. Rosenbluth, A.H. Teller, and E. Teller. Equations of state calculations by fast computing machines. *Journal of Chemical Physics*, 21:1087–1092, 1953.

[12] H. Omre and K. B. Halvorsen. The Bayesian bridge between simple and universal kriging. *Mathematical Geology*, 21(7):767—786, 1989.

[13] H. Omre, Knut Sølna, and H. Tjelmeland. Simulation of random functions on large lattices. Technical report, Norwegian Computing Center, 1991. NR-notat SAND/04/91.

[14] G. J. G. Upton and B. Fingleton. *Spatial Data Analysis by Example, Volume 2.* Wiley, NY, 1989.

DISCUSSION FOLLOWING SESSION NO 7

Chairman: Colin Daly

Papers: Galli; Georgsen

Jeulin	You have a very nice but somewhat complex model. It seems that there are parameters which do lead to flexibility but I cannot see how you set up the model in practice.
Georgsen	There are problems because of the number of parameters. We hardly know how they influence the model and it is even harder for the users. By looking at the realizations, one can tune the parameter values. That is what users in oil companies do.
Omre	I agree that there are many parameters but some of the objectives in using marked point process is that the meaning of some of the parameters is intuitively obvious. For example; size, depth, width... We are also formulating this in a Bayesian setting; users can input prior guesses on these parameters. So I agree that the number of parameters is large but it is not such a big problem when the users have an intuitive understanding of what the parameters are.
Georgsen	The users have a good idea of the expected width and the expected thickness, and perhaps the standard deviation. The correlations are hard to assess. What happens if you set the correlation between width and thickness to 0.5 rather than 0.8? Nobody can answer that, until they have tried.

The chairman then asked that the discussion be widened out to include both papers, and specifically called for questions and discussion on simulating categorical data.

Haas	You mentioned the difficulties of specifying the connectivity between wells.
Georgsen	We are working with specific objects like channels. It is difficult for users to know the channel they see in one well is the same as another 1 km away. Consequently if we have made them specify the probability of being connected, but that is also quite hard because there are so many possible combinations of wells. For example, if you have 10 or 20 wells, there are so many possibilities to specify probabilities for.

M. Armstrong and P. A. Dowd (eds.), Geostatistical Simulations, 251–255.

Gotway (to Alain Galli). Could you tell us briefly what are the top two pros and the top two cons to using the truncated gaussian method ?

Galli Firstly, you have to have a sequential ordering at least from the statistical point of view. If not, you should not use this method.
We have been using this method for many years for fluvial and fluvio–deltaic reservoirs, and once or twice for carbonate reservoirs. It seems to work quite well for carbonate reservoirs as well. But it depends on your point of view. If you have a mathematical background, you will discuss the properties you expect from the reservoir. For example, if you are thinking of a boolean model, you think in terms of objects. So you should not use this method. But having said that, it is difficult to have a clear idea of the limitations of performance of a particular method for a given field, because you have to carry out the study, then try to do a history match with that and you have to repeat this with other methods. So I think it is very difficult to answer your questions. In our opinion, it is an interesting method but we are nevertheless working on others such as the boolean method. We do not think that we could use this method (or any other) for all reservoirs. Even for a given reservoir, we believe that the method to use depends upon the problem posed. There are some obvious cases in which the gaussian cannot be used. For example, on the work that Hu Linying did with the IFP and Elf, on the Roda sandstone, the problem was clearly different. There were prograding bars and each of these has complex geometry with different correlation lines. Moreover, between the bars were permeability barriers. So in that case, we could not use the truncated gaussian method, or any of the indicator methods.

Jeulin With the truncated gaussian method, you can set a series of thresholds to get a multi–component simulation that corresponds to nested facies. Have you had any applications where you have used more thresholds than was required for the typical cases ?

Galli No, because the problem is to decide to what interval each value should be assigned. It is mainly a problem for the conditioning.
I forgot to mention something in my answer to Carol Gotway. When we went to use this method we start with a classical statistical analysis, to see if there is an ordering relation and to find out what relations can be seen between the different facies. For example, there have been cases where this showed that the method would not work. Jacques Rivoirard presented some examples of this at the Troia congress. It was clear that

there were cases where this method could work and others where it would not.

Haas — You have explained the advantages of the method. We should distinguish several cases. Firstly when there are only two lithofacies. In that case, I do not see why one would use an underlying gaussian function. For example, if you compare that with an indicator function method.

Galli — Our policy is to use consistent models, and...

Haas — But you have only 1 variogram in that case.

Galli — The problem is to choose a variogram model that can represent an indicator function, and that is not so easy.

Haas — O.K.

Galli — Even with two lithotypes, we see strong vertical variations, at least. With indicators, you are able to include this in your evaluations, as was shown by Valérie Langlais. The advantage of the truncated gaussian is that you are able to calibrate your parameters (i.e. your variogram). With the indicator method, if you include this trend you cannot have a good idea of the vertical variogram, because you are not in a stationary case. So generally you will overestimate the vertical continuity.

Haas — The general case with several lithofacies is more interesting. You think that there is an ordering, and an underlying physical function. So in that case, you can simulate this function directly. Moreover you can have some indirect control like the log of the sand/shale ratio which is continuous. So we can simulate.

Galli — The main point is that we believe that we should split the work into two steps : first taking the geology into account. Behind this method, is the general idea that heterogeneity is controlled at least partly by the sedimentological process. That is, we first simulate the sedimentology and then in a second step, we simulate porosities and permeabilities. It is better to start with that. For example in this case *(he shows a transparencies)* by doing it that way ; we see that this reservoir has to be divided into three parts. Two of them are marine parts with entirely different characteristics.

Haas — You have this information on your well logs.

Galli — It is not easy to distinguish that from logs. You have to make a sedimentological analysis.

Haas Generally geologists make these truncations themselves but they have the information at the beginning. You lose a lot if you try to imagine this function without the original function.

Omre In our discussions with geologists, they very often consider that there is a hierarchy in the processes. There is one driving process, and some others that are conditioned on that. In one of our papers you can see that we have presented a hierarchical system. *(Ed: the reference is Clemetsen, Hurst, Knaned & Omre, 1990, Computer Program for Evaluation of Fluvial Reservoirs; in North Sea Oil & Gas Reservoirs II, Ed Buller et al, Ceraham & Trotman, pp 373–387)* We have rivers, and from the rivers something happens, and so forth. For us, it is important to have building blocks like marked point processes to construct this model, and then try to find suitable simulation algorithms. We are on our way to doing that. You had a good idea using 2 dimensions, and instead of truncating in only 1 dimension but having 2 or higher dimensions so that you can get hierarchical results. Have you tested that in reality ? You just sketched...

Galli I carried out a small application a long time ago (may be in 1988) just discussing with the geologists, the sedimentologists from the IFP.
They were working on Long Nab, that outcrop in Yorkshire, and were trying not only to look at the interpretation but also to draw the channels. The geologist was quite puzzled because on the upper part of the outcrop he could identify isolated channels. In fact, five were seen on the western side. But he could not draw them because he knew that there were more than five, but they were small. At the wells he knew where they were, but he could not draw them elsewhere. He explained that the flood plain had been been eroded by channels.

Omre We are very eager to see some results along those lines because I think it is very fruitful. We also have the experience that once one can establish the lithofacies units, you can dig into them and you have upward sorting. So it is important to know the geological units, because at the borders some particular things happen. So seeing only the overlap is not sufficient, you also have to have the initial building blocks. The challenge in the future coming to us from the geologists is to better incorporate this information in our models.

Galli The method is quite interesting because one can take account of complex objects. The main problem with it was how to get accurate estimates of the parameters. I will give you a typical example that occurred in an outcrop in Spain. It was a 3D outcrop ; and they had drilled wells. You could see a particular channel crossing the outcrop on two sides, only 300 m apart. They decided to drill an extra well to check the channel's position in between. But unfortunately the channel was not where they had expected it. That is, given sightings only 300 m apart and a lot of general information on its shape, you cannot predict its location accurately.

Omre That is why we are using stochastic models.

Galli Yes but you have to input some idea of its parameters.

Omre We have not solved all problems. We are talking about concepts, different concepts for doing this. So for us, it is important to have building blocks. Conditioning is very hard for us. Simulating is simple but conditioning is hard on the branches.

Jeulin An additional comment : the idea is to simulate nested random sets. A long time ago (around 1979) we wanted to develop models for complex petrographical facies at the microscopic scale. These are not geological facies but it is the same problem. We developed models like this not using the properties of the gaussian distribution, but using boolean models with different grains with sequence sets (a first set then a second outside and so on). It is very similar to the approach described by Alain Galli but for a different problem. It was very efficient for simulating complex facies. So why not use it on a very large scale such as this ?

The chairman then thanked the two speakers for very interesting papers, and then he closed the workshop.

Quantitative Geology and Geostatistics

1. F. M. Gradstein, F. P. Agterberg, J. C. Brower and W. S. Schwarzacher: *Quantitative Stratigraphy*. 1985 ISBN 90-277-2116-5
2. G. Matheron and M. Armstrong (Eds.): *Geostatistical Case Studies*. 1987 ISBN 1-55608-019-0
3. *Cancelled*
4. M. Armstrong (Ed.): *Geostatistics*. Proceedings of the 3rd International Geostatistics Congress, held in Avignon, France (1988), 2 volumes. 1989 Set ISBN 0-7923-0204-4
5. A. Soares (Ed.): *Geostatistics Tróia '92*, 2 volumes. 1993 Set ISBN 0-7923-2157-X
6. R. Dimitrakopoulos (Ed.): *Geostatistics for the Next Century*. 1994 ISBN 0-7923-2650-4
7. M. Armstrong and P.A. Dowd (Eds.): *Geostatistical Simulations*. 1994 ISBN 0-7923-2732-2

MIX
Papier aus verantwortungsvollen Quellen
Paper from responsible sources
FSC® C105338

If you have any concerns about our products,
you can contact us on
ProductSafety@springernature.com

In case Publisher is established outside the EU,
the EU authorized representative is:
Springer Nature Customer Service Center GmbH
Europaplatz 3, 69115 Heidelberg, Germany

Printed by Libri Plureos GmbH
in Hamburg, Germany